Transcriptional Control of Neural Crest Development

Colloquium Series on Developmental Biology

Editors
Daniel S. Kessler, *University of Pennsylvania School of Medicine*

Developmental biology is in a period of extraordinary discovery and research in this field will have a broad impact on the biomedical sciences in the coming decades. Developmental Biology is interdisciplinary and involves the application of techniques and concepts from genetics, molecular biology, biochemistry, cell biology, and embryology to attack and understand complex developmental mechanisms in plants and animals, from fertilization to aging. Many of the same genes that regulate developmental processes underlie human regulatory gene disorders such as cancer and serve as the genetic basis of common human birth defects. An understanding of fundamental mechanisms of development is providing a basis for the design of gene and cellular therapies for the treatment of many human diseases. Of particular interest is the identification and study of stem cell populations, both natural and induced, which is opening new avenues of research in development, disease, and regenerative medicine. This eBook series is dedicated to providing mechanistic and conceptual insight into the broad field of Developmental Biology. Each issue is intended to be of value to students, scientists and clinicians in the biomedical sciences.

Copyright © 2010 by Morgan & Claypool Life Sciences

All rights reserved. No part of this publication may be reproduced, stored in a retrieval system, or transmitted in any form or by any means—electronic, mechanical, photocopy, recording, or any other except for brief quotations in printed reviews, without the prior permission of the publisher.

Transcriptional Control of Neural Crest Development
Brian L. Nelms and Patricia A. Labosky
www.morganclaypool.com

ISBN: 9781615040483 paperback

ISBN: 9781615040490 ebook

DOI: 10.4199/C00010ED1V01Y201003DEB001

A Publication in the Morgan & Claypool Life Sciences Publishers series

DEVELOPMENTAL BIOLOGY

Book #1

Series Editor: Daniel S. Kessler, Ph.D., University of Pennsylvania

Series ISSN Pending

Transcriptional Control of Neural Crest Development

Brian L. Nelms and Patricia A. Labosky
Vanderbilt University

DEVELOPMENTAL BIOLOGY #1

MORGAN & CLAYPOOL LIFE SCIENCES

ABSTRACT

The neural crest is a remarkable embryonic population of cells found only in vertebrates and has the potential to give rise to many different cell types contributing throughout the body. These derivatives range from the mesenchymal bone and cartilage comprising the facial skeleton, to neuronal derivatives of the peripheral sensory and autonomic nervous systems, to melanocytes throughout the body, and to smooth muscle of the great arteries of the heart. For these cells to correctly progress from an unspecified, nonmigratory population to a wide array of dynamic, differentiated cell types—some of which retain stem cell characteristics presumably to replenish these derivatives—requires a complex network of molecular switches to control the gene programs giving these cells their defining structural, enzymatic, migratory, and signaling capacities. This review will bring together current knowledge of neural crest-specific transcription factors governing these progressions throughout the course of development. A more thorough understanding of the mechanisms of transcriptional control in differentiation will aid in strategies designed to push undifferentiated cells toward a particular lineage, and unraveling these processes will help toward reprogramming cells from a differentiated to a more naive state.

KEYWORDS

neural crest, transcription, regulation, differentiation, Pax, Fox, Sox, Hox, bHLH, zinc finger, homeobox

Contents

CHAPTER 1

Introduction

1.1 THE ROLE OF TRANSCRIPTIONAL REGULATION IN DEVELOPMENT

During development, an organism that starts as a single cell rapidly progresses to a dynamic collection of cells dividing and differentiating into the multiple lineages that make up the diverse systems of the body plan. All cells of a given organism contain the same DNA, but it is how information is selectively read from this master code that directs these cells to divide and differentiate myriad of cell types throughout the body. In combination with sets of maternal and zygotic signaling molecules that initiate regulatory cascades, it is transcription factors that act as switches to turn on or off gene expression or even modulate the precise expression output of a gene, functioning more like rheostats to control proper transcript levels rather than simply an all-or-none response. Combinatorial control by multiple transcription factors working in concert can also confer cell type-specific regulation of target genes to produce specialized cell types. Interactions between multiple transcription factors at shared target genes result in complex gene regulatory networks, but understanding and unraveling these networks will aid in understanding how different cell types arise and perhaps further the ability to direct cells to adopt a given fate. This review will focus on transcriptional regulation specifically within the neural crest (NC) lineage, keeping in mind the broad potential of these cells for differentiation into many cell types, with an aim of compiling and clarifying the roles of transcription factors across the many diverse NC derivatives (Figure 1.1).

1.2 NEURAL CREST CELLS HAVE REMARKABLE DEVELOPMENTAL PLASTICITY

The NC is one distinguishing feature of vertebrates (the studies covered here will mostly include humans and the model organisms mouse, chicken, frog, and zebrafish), and regulation of the NC is highly conserved among these different species. Many of the features setting vertebrates apart from invertebrates—skin pigmentation, jaw development, an enlarged skull for greater brain capacity, and the autonomic nervous system—are made up of cell types derived from the NC. Neural crest cells (NCCs) are specified at the region of the neural plate border around the time of neurulation and when the neural tube begins to fold, and perhaps even earlier. Gradients of signaling from the

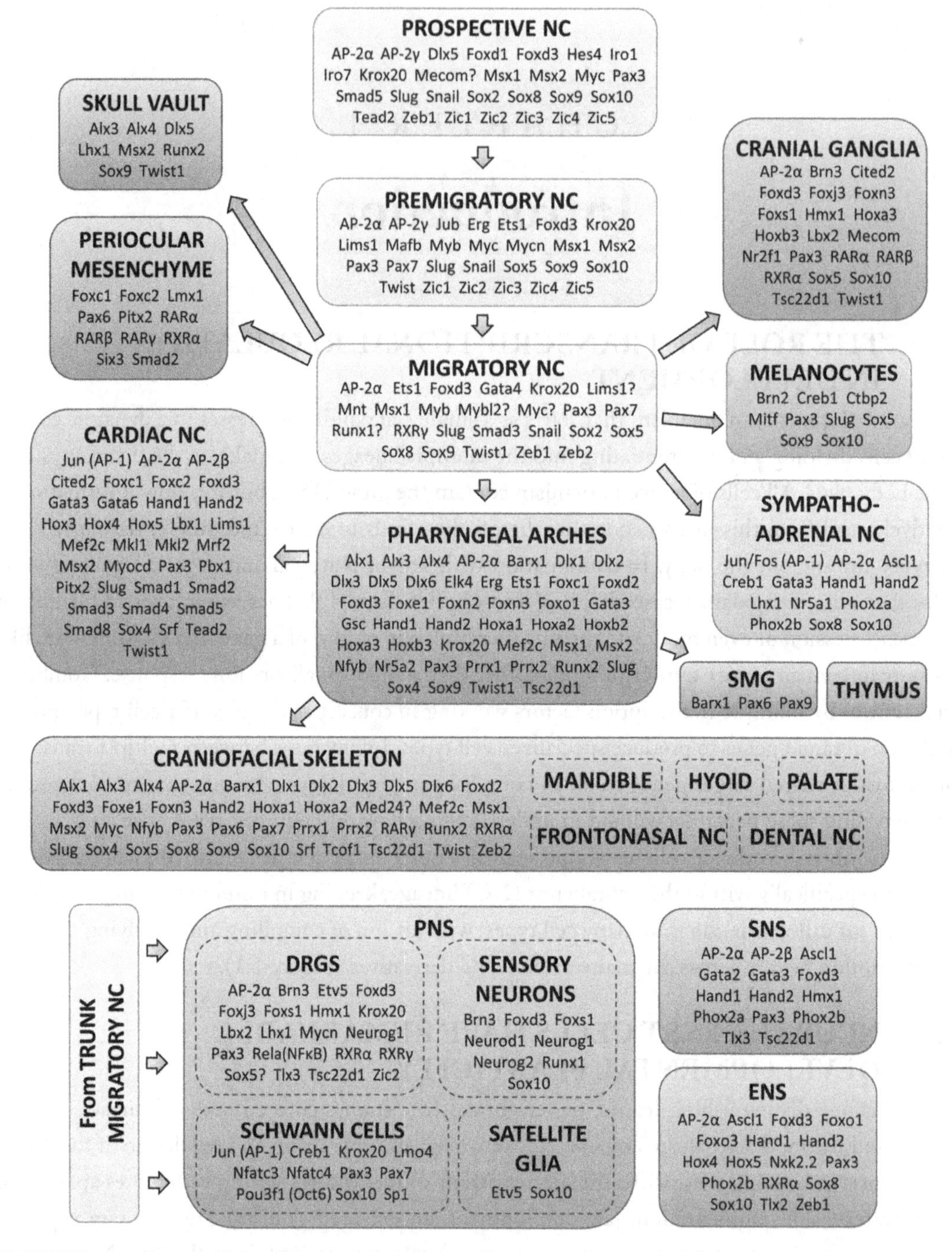

FIGURE 1.1: Distribution of transcription factors known to play a role in developing neural crest lineages. For references, refer to text. Boxes are color-coded based on cell type: green, early NC; purple, mesenchymal NC; blue, neural/glial NC; red, cardiac NC; orange, pigment cells; aquamarine, sympathoadrenal NC. Abbreviations are as follows: NC, neural crest; DRGs, dorsal root ganglia; ENS, enteric nervous system; PNS, peripheral nervous system; SMG, submandibular gland; SNS, sympathetic nervous system.

BMP, Wnt, and FGF signaling pathways, both in the ectoderm and from surrounding tissues, are key components of this process (for detailed reviews of NC induction/specification, see Basch and Bronner-Fraser, 2006; Sauka-Spengler and Bronner-Fraser, 2008; Steventon et al., 2005). After NCCs are specified, they begin to migrate out from the forming dorsal neural tube and traverse one of a number of distinct pathways. The anterior-to-posterior level of origin of these NCCs and the path of migration that the NCCs take are important determinants of cell fate, but even NCCs from regions that do not normally give rise to a specific cell type can often do so if transplanted elsewhere (for detailed reviews of NC migration and fate potential, see Dupin et al., 2006; Harris and Erickson, 2007; Kuriyama and Mayor, 2008). The NC is a population with much initial plasticity, so much so that it has been termed by some as the fourth germ layer. Although the NC arises from ectoderm, it can contribute cells to diverse structures such as the facial skeleton, teeth, peripheral nervous system, great arteries of the heart, adrenal cells, and more. Also, an individual NCC can give rise to multiple NC derivatives, indicating a multipotent state (Dupin et al., 2006, Dupin et al., 2007; Le Douarin et al., 2008). Understanding the basic biology of this population is compelling for many reasons, but especially with regard to understanding the control of differentiation programs.

NCCs arise from anterior-to-posterior levels that can be divided into cranial, vagal, trunk, and sacral NC regions. These locations are generally predictive of the fates that these NCCs adopt during normal development. The cranial NC, which is largely aligned with the developing midbrain/hindbrain region of neural tissue, gives rise to melanocytes, cranial ganglia, connective tissue of the head and neck, the cartilage and bone of the craniofacial skeleton, the skull vault, the dentine of the teeth, bones of the inner ear, the cornea and sclera of the eyes, ciliary and eye attachment muscles, and the parafollicular cells of the thyroid. Cells from the vagal NC (from the mid-otic placode to somite 7) give rise to enteric and mesenteric ganglia, celiac ganglia, and aortical renal ganglia. The cardiac NC is contained within the vagal NC region, and cardiac NCCs normally become smooth muscle cells of the great arteries and the aorticopulmonary septum, and pericytes that are associated with the great arteries. These cells also perform a morphogenetic function; they are responsible for septation of the outflow tract. NCCs of the trunk region form pigment cells, dorsal root ganglia, sensory sympathetic ganglia, and chromaffin cells and epinephrine-producing cells of the adrenal medulla. The sacral NC gives rise to enteric ganglia and the parasympathetic ganglia associated with the alimentary canal and blood vessels (for in-depth review of NC lineages and derivatives, see Dupin et al., 2006).

1.3 THE ROLE OF TRANSCRIPTION FACTORS IN NEURAL CREST STEM CELLS AND INDUCED PLURIPOTENT STEM CELLS

Although most of the NCCs undergo progressive differentiation, resulting in mostly terminally differentiated, specialized cell types, there are subsets of cells in different lineages that somehow do not undergo terminal differentiation, or perhaps are maintained within niches that allow fate changes when necessary, further highlighting the developmental plasticity of this population. Based mostly on mammalian studies (mouse, rat, and human), a number of adult NC-derived cells have been identified as cells that retain stem cell properties. These include (1) skin-derived precursors (Fernandes and Miller, 2009; Toma et al., 2005) that are NC-derived cells in the dermis that persist in a dermal papilla niche, (2) epidermally derived EPI-NCSCs (Sieber-Blum and Hu, 2008; Sieber-Blum et al., 2006) that reside in the bulge region of the hair follicle, (3) boundary cap NC-derived stem cells (Aldskogius et al., 2009), (4) adult palatum-derived NCSCs (Widera et al., 2009), (5) corneal limbus-derived NCSCs (Brandl et al., 2009), and (6) a few adult tooth-derived NCSC-like populations (Coura et al., 2008; Techawattanawisal et al., 2007) (for further review, see Le Douarin et al., 2008; Teng and Labosky, 2006). These NCSC populations often retain an expression profile consisting of transcription factors characteristic of both embryonic stem cells and early NC. In addition to the study of NC-derived stem cell populations, in the past few years, a number of important breakthroughs have been made in the generation of induced pluripotent stem cells (iPSCs) (Park et al., 2008; Takahashi and Yamanaka, 2006; Takahashi et al., 2007; Yu et al., 2009). iPSCs can be derived by the expression of just a few key transcription factors capable of reprogramming differentiated somatic cells into a more immature, pluripotent state. Using similar strategies, one handful of transcription factors has also been shown to convert pancreatic exocrine cells into β-like cells in the pancreas (Zhou et al., 2008b), whereas another small cohort can convert noncardiogenic mesoderm into beating cardiomyocytes (Takeuchi and Bruneau, 2009). These studies suggest, in several different lineages, that moving from one cell type to another may be at least closely approximated by the control of only a few master transcription factors, and despite the known importance of signaling cascades, direct manipulation of the transcription factors themselves allows more precise control of multiple signaling pathways.

1.4 NC MISREGULATION IN DISEASE STATES

Because of the NC's extensive role in many different tissue types, defects in these cells are often part of many of the most common congenital defects. Some of the more common disorders that arise from dysfunction of NC development are Waardenburg syndrome (WS), velocardiofacial syndrome (VCFS)/ DiGeorge syndrome, Hirschsprung's Disease, and a large number of congenital heart defects (CHDs) (Pierpont et al., 2007). VCFS has a spectrum of palatal, facial, and heart anomalies (www.nidcd.nih.gov), and patients with DiGeorge syndrome/22q11 deletion syndrome present with similar characteristics. Identification of mutations in human genes highly associated with specific neurocristopathies has added to a more molecular and mechanistic understanding into how normal development proceeds under the control of transcription factors. Misregulation of NC development, or perhaps reacquisition of NCSC-like properties by differentiated cells from the NC lineage, can lead to various cancers of NCC origin: melanoma, neuroblastoma, Ewing's family tumors, primitive peripheral neuroectodermal tumors, schwannomas, pheochromocytomas, neurofibromas, carcinoid tumors, and medullary carcinoma of the thyroid. The importance of a regulated differentiation program under the tight control of transcription factors is highlighted in the prevalence of NC-related tumors such as melanoma and neuroblastoma. In addition to initiating expression of specific transcription factors that regulate development of a differentiated cell type, repressing genes in differentiated lineages is also important. Many cancers arise when silenced transcription factors are reactivated. Perhaps not surprisingly, many identified tumor-causing transcription factors are critical players in the gene regulatory networks involved in normal NC development.

In addition to studying human disease, our broad knowledge of how NC contributes to the developing body plan was assembled from a number of different experimental approaches in model organisms: NC ablation studies (Donnell et al., 2005; Kirby, 1990; Porras and Brown, 2008), chicken–quail chimeras (Le Douarin, 1980; Sieber-Blum et al., 1993), fate-mapping with vital dye injections, Cre-loxP-mediated lineage mapping in mice, and characterization of many NC mutants using transgenic and gene-targeted mouse models and siRNA or morpholino knockdown experiments in cultured cells, zebrafish, or *Xenopus laevis*.

1.5 REVIEWING CURRENT KNOWLEDGE OF NC TRANSCRIPTIONAL CONTROL

In this review, we present an overview of current knowledge on transcription factors expressed in the developing NC and their role in NC induction, migration, and particularly during the course of differentiation. Many more transcription factors than have been studied are undoubtedly involved in NC differentiation. At an increasing pace, more information about the role that both established and novel transcription factors play in the NC is being discovered. Recent and future progress through high throughput screens (microarrays, ChIP on chip, ChIPSeq, other deep-sequencing analyses, etc.) will uncover many more factors and shed light on how they fit into regulatory networks, but much work will be needed to study the precise mechanisms by which they act. Transcription factors may function as repressors, activators, or both if they depend on the recruitment of binding partners to confer activation or repression, and many transcription factors converge on a shared target to act synergistically to affect gene activation or repression. We will focus on interactions between well-described NC transcription factors at both the protein–protein levels and the shared target levels. Another large component of transcriptional regulation is the modification of chromatin by epigenetic marks. In both the iPSC studies and the fate conversion of exocrine to endocrine lineages in the pancreas discussed above, chromatin modifications, in addition to the introduction of transcription factors, are important for these fate changes. Chromatin-associated proteins or inhibitors (e.g., HDAC inhibitors) can often enhance this process, and understanding methylation profiles and histone modifications is important for a full understanding of these induced changes. However, the field of chromatin control of NC properties is relatively untouched, with only a handful of studies directed at NC-specific perturbation of chromatin modifier function (Haberland et al., 2009; Ignatius et al., 2008). A role for a few chromatin-associated factors in NC development has been uncovered, principally as proteins recruited by NC transcription factors as a part of the gene regulatory mechanism affected by perturbation of the specific transcription factor. These studies will not be covered in this review, but this is clearly an exciting and open avenue of research that will surely expand rapidly over the next few years. Many chromatin-associated factors are broadly or ubiquitously expressed, but their activity within the NC has not yet been investigated. On the horizon, multiple approaches, particularly reverse genetics, generating transgenic mice, conditional knockouts, RNAi knockdown, and biochemical assays, should help to further describe the importance of chromatin modifications in the NC.

This review will not serve as an exhaustive compendium of all transcription factors within the neural crest but represents a summary of current knowledge of transcription factor expression and function within the NC and its derivatives with a particular emphasis on development. This summary includes multiple vertebrate species, but for simplicity, the mouse nomenclature and styling for a particular gene will be used with few exceptions. Table 1.1 includes a column listing alternative gene names for a given factor when significant differences occur. Table 1.2 lists all abbreviations used. Where known, the phenotype of mutants, morphants, overexpression data, and so on, will be summarized.

TABLE 1.1: Summary of expression domains and functions of neural crest transcription factors.

FAMILY	TF	SPECIES (ALTERNATIVE NAMES)	KNOWN NC EXPRESSION DOMAINS	KNOWN NC FUNCTIONS	MGI NO.
AP	Fos (AP-1)	Mm, Xl, Rn, Bp (c-Fos)	Early NC, cardiac NC, Schwann cells, chrommafin cells	Cardiac NC outgrowth, Zic induction in early NC	95574
	Jun (AP-1)	Mm, Xl, Rn, Bp (c-Jun)	Early NC, cardiac NC, Schwann cells, chrommafin cells	Cardiac NC outgrowth, Zic induction in early NC	96646
	AP-2α	Mm, Xl, Dr, Gg (Tcfap2a, tfap2a)	Cranial NCCs, migrating cardiac NC	NCC specification, survival	104671
	AP-2β	Mm, Hs, Xl, Dr (Tcfap2b, tfap2b)	NCCs, facial mesenchyme, sympathetic ganglia, adrenal medulla, ductus arteriosus	NCC specification, survival, skeletal differentiation, neuronal gene regulation	104672
	AP-2γ	Mm, Xl, Dr (Tcfap2c, tfap2c)	Premigratory NCCs, migratory NCCs, facial mesenchyme	NC induction	106032
bHLH	Ascl1	Mm, Hs, Dr, Xl, Gg (Mash1, Hash1)	NC progenitors, peripheral neurons	Promotion of neurogenesis	96919

TABLE 1.1 (*continued*)

FAMILY	TF	SPECIES (ALTERNATIVE NAMES)	KNOWN NC EXPRESSION DOMAINS	KNOWN NC FUNCTIONS	MGI NO.
bHLH (*cont'd*)	Hand1	Mm, Gg, Xl, Hs (eHand)	Pharyngeal arches, OFT, autonomic ganglia, enteric ganglia	Mesenchymal NC growth, survival	103577
	Hand2	Mm, Gg, Dr (dHand)	Pharyngeal arches, OFT, sympathetic ganglia	autonomic neurogenesis, inhibition of osteogenesis, cranial NCC survival, cardiac NC formation	103580
	Mitf	Mm, Gg, Hs, Dr, Xl	Melanocytes / pigment cell lineage	Master regulator of melanogenesis	104554
	Mnt	Xl	Migratory NCCs	unknown	109150
	Myc	Mm, Hs, Rn, Xl, Gg, Dr (c-Myc)	Premigratory NC, NCSCs	Maintenance of multipotent progenitor state	97250
	Mycn	Mm, Hs, Gg (n-Myc)	PNS, DRGs, neuroblastoma	Proliferation, neuronal differentiation of progenitors	97357
	Olig3	Mm, Gg, Dr	Prospective NC	Establishing NC-neural plate boundary	2149955
	Tcf3	Mm	NC	Modulating bHLH transcription	98510

bHLH (*cont'd*)	Tcf4	Mm	NC	Modulating bHLH transcription	98506
	Tcf12	Mm	NC	Modulating bHLH transcription	101877
	Tcfe3	Mm	NC	Activating Dct promoter	98511
	Twist1	Mm, Hs, Dr, Xl	Early NC, cranial mesenchyme, pharyngeal arches, OFT, skull vault	Regulation of EMT, regulation of NC mesenchyme proliferation and differentiation	98872
	Usf1	Mm	NC, VSMCs	Control of VSMC gene transcription	99542
	Usf2	Mm	NC, VSMCs	Control of VSMC gene transcription	99961
ETS	Ets1	Mm, Gg, Xl	Premigratory NC, migratory NC, pharyngeal arches (?), melanocyte precursors	Regulation of adhesion molecules, motility, cell specification	95455
	Ets2	Mm	Tooth buds?	Unknown	95456
	Elk1	Mm	NC VSMC precursors	Differentiation of VSMCs	101833

continued on next page

TABLE 1.1 (*continued*)

FAMILY	TF	SPECIES (ALTERNATIVE NAMES)	KNOWN NC EXPRESSION DOMAINS	KNOWN NC FUNCTIONS	MGI NO.
ETS (*cont'd*)	Elk4	Xl (xSap1)	Pharyngeal arches?	Unknown	102853
	Erg	Mm, Hs, Gg, Hs	Pharyngeal arches	EMT, migration, differentiation	95415
	Etv1	Mm, Hs (ER81), Xl	NCCs	Unknown	99254
	Etv4	Dr (Pea3)	NCCs	Unknown	99423
	Etv5	Mm, Rn (Erm)	Multipotent NC, satellite glia, neurons of DRG, peripheral neurons	Control of glial proliferation	1096867
	Fli1	Mm, Hs, Gg, Xl	Mesenchymal NC	Motility, adhesion, ECM regulation	95554
Fox	Foxc1	Mm, Hs, Xl	Pharyngeal arches, periocular mesenchyme, cardiac NC	Specification of peri-ocular mesenchyme progenitors	1347466
	Foxc2	Mm	Cranial NC, cardiac NC	Cardiac NC survival	1347481
	Foxd1	Xl	Premigratory NC	Negative regulation of NC migration	1347463

Fox (*cont'd*)	Foxd2	Xl	Cranial NC, pharyngeal arches,	Unknown	1347471
	Foxd3	Mm, Hs, Gg, Dr, Xl	Premigratory NC, migratory NC,	NC survival, multipotency	1347473
	Foxe1	Hs, Dr	Pharyngeal arches, skeleton, thyroid	Chondrocyte differentiation, proliferation in pharyngeal skeleton	1353500
	Foxf1	Mm, Xl	Palatal/oral mesenchyme	Unknown	1347470
	Foxf2	Mm	Palatal/oral mesenchyme	Unknown	1347479
	Foxj3	Mm	NC, cranial ganglia	Unknown	2443432
	Foxn2	Xl	Premigratory NC, pharyngeal arches	Unknown	1347478
	Foxn3	Xl	Cranial ganglia, pharyngeal arches, mandible	NC differentiation	1918625
	Foxo1	Mm, Xl	Cranial NC, pharyngeal arches, ENS	ENS precursor survival, neurite extension	1890077
	Foxo3	Mm, Xl	Cranial NC, pharyngeal arches, ENS	ENS precursor survival, neurite extension	1890081
	Foxs1	Mm	Cranial ganglia, peripheral ganglia, DRG, enteric ganglia	NC differentiation toward sensory neuron lineage	95546

continued on next page

TABLE 1.1 (*continued*)

FAMILY	TF	SPECIES (ALTERNATIVE NAMES)	KNOWN NC EXPRESSION DOMAINS	KNOWN NC FUNCTIONS	MGI NO.
Homeo-box	Alx1	Mm	Craniofacial cartilage	Craniofacial patterning, cell survival	104621
	Alx3	Mm	Craniofacial cartilage	Craniofacial patterning, cell survival	1277097
	Alx4	Mm	Craniofacial cartilage	Craniofacial patterning, cell survival	108359
	Barx1	Mm, Hs, Dr	First and second pharyngeal arch mesenchyme	Osteochondrogenic progenitor proliferation, differentiation	103124
	Dlx1	Mm, Dr	Migratory cranial NC, teeth, mandible	Specification of odontogenic NC	94901
	Dlx2	Mm, Dr	Migratory cranial NC, teeth, mandible	Specification of odontogenic NC	94902
	Dlx3	Mm, Dr	Migratory cranial NC, teeth, mandible	Specification of odontogenic NC	94903
	Dlx5	Mm, Dr	Migratory cranial NC, teeth, mandible	Specification of odontogenic NC	101926
	Dlx6	Mm, Dr	Migratory cranial NC, teeth, mandible	Specification of odontogenic NC	101927

Homeo-box *(cont'd)*	Emx1	Mm	Non-melanocyte NC	Negative regulation of Mitf	95387
	Emx2	Mm	Non-melanocyte NC	Negative regulation of Mitf	95388
	Gsc	Mm, Dr	Pharyngeal arches, mandible, middle ear	NC patterning and morphogenesis	95841
	Gsc2	Hs	NC	Unknown	892006
	Hlx	Mm	ENS	Unknown	96109
	Hmx1	Mm	DRGs, sympathetic ganglia	Unknown	107178
	Lbx1	Mm	Cardiac NC	Specification of cardiac NC, Pax3 regulation	104867
	Lbx2	Mm	Cardiac NC, DRGs, cranial ganglia	Specification of cardiac NC, Pax3 regulation	1342288
	Msx1	Mm, Dr	Prospective NC, pharyngeal arches, craniofacial skeleton	Mesenchymal–epithelial interactions, NC induction	97168
	Msx2	Mm	Prospective NC, pharyngeal arches, craniofacial skeleton	Mesenchymal-epithelial interactions, NC induction	97169
	Phox2a	Mm	Sympathoadrenal lineage	Sympathetic neuron specification	106633

continued on next page

TABLE 1.1 *(continued)*

FAMILY	TF	SPECIES (ALTERNATIVE NAMES)	KNOWN NC EXPRESSION DOMAINS	KNOWN NC FUNCTIONS	MGI NO.
Homeobox ***(cont'd)***	Phox2b	Mm, Hs	Sympathodrenal lineage, sympathetic ganglia, enteric ganglia	Sympathetic neuron specification	1100882
	Pitx2	Mm	Periocular mesenchyme, cardiac NC	Specification of periocular mesenchyme derivatives	109340
	Prrx1	Mm	Facial mesenchyme, middle ear	Epithelial-mesenchymal interactions	97712
	Prrx2	Mm	Pharyngeal arches, tooth mesenchyme	Epithelial-mesenchymal interactions	98218
	Runx1	Mm, Gg	Migratory NCCs, sensory neurons	Sensory neuron fate specification	99852
	Runx2	Mm, Gg, Dr	Pharyngeal arches, mandible, skull	Regulation of osteogenesis	99829
	Tlx1	Mm, Gg	Cranial NC	Differentiation, maintenance of neuronal progenitors	98769

Homeo-box (*cont'd*)	Tlx2	Mm	ANS, ENS	Differentiation, maintenance of ANS neuronal progenitors	1350935
	Tlx3	Mm, Gg	ANS sympathetic ganglia, DRG	Differentiation, maintenance of neuronal progenitors	1351209
Hox	Hoxa1	Mm, Xl	cranial NCCs, pharyngeal arches, craniofacial skeleton, cartilage precursors	Patterning cranial NCCs, segmentation, RA response	96170
	Hoxb1	Mm, Gg, Xl	cranial NCCs, pharyngeal arches, craniofacial skeleton, cartilage precursors	Patterning cranial NCCs, segmentation, RA response	96182
	Hoxd1	Xl	cranial NCCs, pharyngeal arches, craniofacial skeleton, cartilage precursors	Patterning cranial NCCs, segmentation, RA response	96201
	Hoxa2	Mm, Xl, Dr	Migrating cranial NCCs, cartilage precursors	Patterning cranial NCCs, segmentation, RA response	96174
	Hoxb2	Mm, Xl, Dr	Migrating cranial NCCs, cartilage precursors	Patterning cranial NCCs, segmentation, RA response	96183

continued on next page

TABLE 1.1 *(continued)*

FAMILY	TF	SPECIES (ALTERNATIVE NAMES)	KNOWN NC EXPRESSION DOMAINS	KNOWN NC FUNCTIONS	MGI NO.
Hox *(cont'd)*	Hoxa3	Mm	Neuronal and mesenchymal NC derivatives, pharyngeal arches, cardiac NC, thymus	Patterning cranial NCCs, segmentation, RA response	96175
	Hoxb3	Mm	Neuronal and mesenchymal NC derivatives, pharyngeal arches, cardiac NC, thymus	Patterning cranial NCCs, segmentation, RA response	96184
	Hoxd3	Mm	Neuronal and mesenchymal NC derivatives, pharyngeal arches, cardiac NC, thymus	Patterning cranial NCCs, segmentation, RA response	96207
	Hoxa4	Mm, Hs	Vagal NCCs, ENS	Patterning	96176
	Hoxb4	Mm, Hs	Vagal NCCs, ENS	Patterning	96185
	Hoxc4	Mm, Hs	Vagal NCCs, ENS	Patterning	96195
	Hoxd4	Mm, Hs	Vagal NCCs, ENS	Patterning	96208
	Hoxa5	Mm, Hs	Vagal NCCs	Patterning	96177
	Hoxb5	Mm, Hs	Vagal NCCs	Patterning	96186
	Hoxc5	Mm, Hs	Vagal NCCs	Patterning	96196
	Hoxb6	Mm	Trunk NCCs	Patterning	96187

Hox (*cont'd*)	Hoxa7	Mm	Trunk NCCs	Patterning	96179
	Hoxb7	Mm	Trunk NCCs	Patterning	96188
	Hoxb8	Mm	Trunk NCCs	Patterning	96189
	Hoxb9	Mm	Trunk NCCs	Patterning	96190
	Hox10-13	N/A	Sacral NCCs	Patterning	N/A
Hox cofactors	Meis1	Mm, Xl, Dr	Pharyngeal arches	Mm, Dr, Xl	104717
	Pbx1	Mm, Xl	Pharyngeal arches, OFT, adrenal gland	Hox cofactor, regulation of Pax3	97495
	Pknox1	Dr (Prep1)	Pharyngeal arches	Hox cofactor, NC fate specification, craniofacial chondrogenesis	1201409
LIM	Jub	Xl	Premigratory NC	NC EMT, corepressor of Snail and Slug	1341886
	Lhx1	Mm, Xl (Lim1)	DRGs, adrenal gland, skull vault	Unknown	99783
	Lhx7	Mm	NCCs	Unknown	1096343
	Lims1	Mm (Pinch)	Cardiac NC	NCC adhesion, migration	1195263

continued on next page

TABLE 1.1 *(continued)*

FAMILY	TF	SPECIES (ALTERNATIVE NAMES)	KNOWN NC EXPRESSION DOMAINS	KNOWN NC FUNCTIONS	MGI NO.
LIM ***(cont'd)***	Lmo4	Mm	Schwann cell precursors	Mediating transcription factor interactions	109360
	Pdlim5	Mm (Enh)	PNS, neuroblastoma	Id regulation during neuronal differentiation	1927489
Pax	Pax3	Mm, Hs, Gg, Dr, Xl	Premigratory NC, migratory NC, melanocytes	NC progenitor proliferation	97487
	Pax6	Mm, Rn, Gg, Xl, Dr	Ocular mesenchyme, submandibular gland mesenchyme	Epithelial-mesenchymal interactions	97490
	Pax7	Mm, Gg, Dr, Xl	Premigratory NC, Migratory NC, melanocytes	NC formation, differentiation	97491
	Pax9	Mm	Cranial NC, pharyngeal arches, tooth mesenchyme, SMG mesenchyme	unknown	97493
POU	Pou2f1	Xl (Oct1)	NCCs	NCC specification, differentiation	101898
	Pou3f1	Mm, Rn (Oct6)	Schwann cell precursors	Regulating Schwann cell transitions	101896

POU (*cont'd*)	Pou3f2	Mm (Brn2/Oct5)	Melanoblasts	Inhibition of melanoblast differentiation	101895
	Pou4f1	Mm, Hs, Gg, Xl (Brn3)	Migratory NCCs, cranial ganglia, DRGs, sensory neurons	Differentiation/survival of NCC-derived sensory neurons	102525
	Pou5f1	Mm, Hs, Rn (Oct4)	NCSCs	Maintenance of pluripotency and self-renewal	101893
RAR/ RXR	RARα	Mm, Hs, Gg, Dr, Xl	Pharyngeal arches, periocular mesenchyme, cranial ganglia, thymus	Craniofacial skeletal patterning, neurogenesis	97856
	RARβ	Mm, Hs, Gg, Dr, Xl	Pharyngeal arches, periocular mesenchyme, cranial ganglia, thymus	Craniofacial skeletal patterning, neurogenesis	97857
	RARγ	Mm, Hs, Gg, Dr, Xl	Pharyngeal arches, periocular mesenchyme, cranial ganglia, thymus	Craniofacial skeletal patterning, neurogenesis	97858
	RXRα	Mm, Hs, Gg, Dr, Xl	Trunk premigratory NCCs, PNS, cranial ganglia	Patterning, neurogenesis	98214
	RXRβ	Mm, Hs, Gg, Dr, Xl	Trunk premigratory NCCs, PNS, cranial ganglia	Patterning, neurogenesis	98215

continued on next page

TABLE 1.1 *(continued)*

FAMILY	TF	SPECIES (ALTERNATIVE NAMES)	KNOWN NC EXPRESSION DOMAINS	KNOWN NC FUNCTIONS	MGI NO.
RAR/ RXR *(cont'd)*	RXRγ	Mm, Hs, Gg, Dr, Xl	Trunk premigratory NCCs, PNS, cranial ganglia	Patterning, neurogenesis	98216
Smad	Smad1	Mm	Prospective NC, dental NC, VSMCs	BMP signal transducer, smooth muscle differentiation	109452
	Smad2	Mm	Dental NC, periocular mes-echyme, VSMCs	TGFβ signal trans-ducer, smooth muscle differentiation	108051
	Smad3	Mm	Dental NC, trunk migratory NCCs, VSMCs	TGFβ signal trans-ducer, smooth muscle differentiation	1201674
	Smad4	Mm	Dental NC, cardiac NC, VSMCs	Common BMP/TGFβ transducer, smooth muscle differentiation	894293
	Smad5	Mm, Dr	Prospective NC, dental NC, VSMCs	BMP signal transducer, smooth muscle differentiation	1328787
	Smad6	Mm	Dental NC	Inhibitory Smad	1336883
	Smad7	Mm	Dental NC	Inhibitory Smad	1100518

Smad (*cont'd*)	Smad8	Mm	NC	BMP signal transducer, smooth muscle differentiation	
	Smad9	Mm	NC	TGFβ signal transducer	1859993
Sox	Sox2	Mm, Hs, Dr, Xl, Gg	Migratory NC, postmigratory NCCs of the PNS, NCSCs	NC induction, maintenance	98364
	Sox4	Mm, Hs, Gg	Pharyngeal arches, craniofacial mesenchyme, OFT	Myofibroblast-endocardial interactions in OFT	98366
	Sox5	Mm, Gg (Lsox5)	Premigratory NC, glia, neurons, melanocytes, craniofacial cartilage	NC induction, chondrogenesis	98367
	Sox8	Mm	Vagal NCCs, enteric glia, adrenal medulla	Maintaining vagal NCSC pool, NC formation, timing of NC specification	98370
	Sox9	Mm, Hs, Gg, Dr, Xl, Rn	Early NC, NCSCs, chondrogenic NC, pharyngeal arches, melanocytes	NC specification, EMT, chondrogenic differentiation and proliferation, NC survival	98371
	Sox10	Mm, Hs, Gg, Xl, Dr	Premigratory NC, multipotent migratory NC, NCSCs, boundary caps, Schwann cells	NC induction, differentiation, glial survival	98358

continued on next page

TABLE 1.1 *(continued)*

FAMILY	TF	SPECIES (ALTERNATIVE NAMES)	KNOWN NC EXPRESSION DOMAINS	KNOWN NC FUNCTIONS	MGI NO.
HMG	Hmga1	Hs	Adrenal gland, neuroblastoma	Unknown	96160
	Hmga2	Hs, Xl	NCC derivatives, neuroblastoma	Unknown	101761
	Lef1	Mm, Gg, Xl	Pharyngeal arches	Fate specification, transduction of Wnt signals	96770
	Tcf7	Mm, Gg, Xl	Pharyngeal arches, adrenal gland	Fate specification, transduction of Wnt signals	98507
ZF	Gata2	Gg, Dr	Sympathetic neuron precursors	Neuronal specification, differentiation	95662
	Gata3	Mm	Sympathetic neuron precursors, OFT, thymus, sympathoadrenal lineage	Regulation of noradren-aline synthesis, regula-tion of cardiovascular transcription	95663
	Gata4	Mm	Migratory NCCs, adrenal medulla, OFT, boundary caps, bcNCSCs, NCSCs	NCSC regulation	95664

ZF (*cont'd*)	Gata6	Mm, Xl	Migratory NCCs, chondrogenic NC, VSMCs	Patterning OFT and arch arteries	107516
	Krox20	Mm, Hs, Rn, Gg, Dr, Xl (Egr2)	R5-associated NCCs, pharyngeal arches, bcNCSCS, Schwann cells, cranial ganglia, DRG	Hindbrain and NCC patterning, Schwann cell transitions	95296
	Mecom	Mm, Hs (Evi1)	PNS-associated NCCs, cranial ganglia	Neuroectodermal differentiation	95457
	Nczf	Mm	NC	unknown	None
	Prdm1	Mm, Dr (Blimp1)	Premigratory NC, pharyngeal arches	NC induction	99655
	Prdm2	Hs (Riz1)	Chromaffin cells	Tumor suppression	107628
	Snail	Mm, Hs, Gg, Dr, Xl (Snail1)	Premigratory NC, migratory NCCs	NC specification, initiation of EMT	98330
	Slug	Mm, Hs, Gg, Dr, Xl, Rn, Quail (Snail2)	Premigratory NC, early migratory NCCs, OFT	NC specification, initiation of EMT	1096393
	Zeb1	Hs, Xl (deltaEF1)	Prospective NC, migratory cranial NC, ENS	Smad corepressor	1344313
	Zeb2	Mm, Xl (SIP1)	Migratory cranial NC, craniofacial skeleton	Smad corepressor	1344407

continued on next page

TABLE 1.1 *(continued)*

FAMILY	TF	SPECIES (ALTERNATIVE NAMES)	KNOWN NC EXPRESSION DOMAINS	KNOWN NC FUNCTIONS	MGI NO.
ZF (*cont'd*)	Zfp704	Mm (Gig1)	Migratory cranial NC	unknown	2180715
	Zic1	Mm, Xl, Gg	Premigratory NC	NC induction	106683
	Zic2	Mm, Hs, Gg, Xl	Premigratory NC	NC induction	106679
	Zic3	Mm, Gg, Xl	Premigratory NC	NC induction	106676
	Zic4	Mm, Gg, Xl	Premigratory NC	NC induction	107201
	Zic5	Mm, Xl	Premigratory NC	NC induction	1929518
Other	Atf1	Hs	Clear cell sarcoma of tendons / malignant melanoma of soft parts	unknown	1298366
	Cited2	Mm, Hs	Cardiac NC, OFT	AP-2 coactivator, Ep300/Crebbp interaction	1306784
	Cited4	Hs	NC	AP-2 coactivator, Ep300/Crebbp interaction	1861694
	Creb1	Mm, Gg	Melanocytes, sympathoadrenal lineage, Schwann cells	Mediation of cAMP signaling	88494

Other (*cont'd*)	Ctbp1	Xl	NC	Zeb1 and Zeb2 corepressor	1201685
	Ctbp2	Mm	Melanocytes	Sox5 corepressor	1201686
	Dawg	Xl	Cranial NC	Unknown	Only Xl
	Mafb	Mm, Dr, Gg (Kreisler)	Cranial NCCs associated with r5 and r6	Hindbrain NCC patterning and specification	104555
	Med12	Dr (Trap230)	NC, craniofacial skeleton	Sox9 coactivator	1926212
	Med24	Dr (Trap100)	Craniofacial skeleton, ENS	Regulation of ENS progenitor proliferation	1344385
	Mef2a	Mm	NCCs	Unknown	99532
	Mef2c	Mm	NCCs, cardiac NC, VSMCs, pharyngeal arches, craniofacial skeleton	Control of vascular smooth muscle gene transcription, pharyngeal skeleton patterning	99458
	Mef2d	Mm	NCCs	Unknown	99533
	Mrf2	Mm	NC, cardiac NC, VSMCs	Control of vascular smooth muscle gene transcription, differentiation	2175912

continued on next page

TABLE 1.1 (*continued*)

FAMILY	TF	SPECIES (ALTERNATIVE NAMES)	KNOWN NC EXPRESSION DOMAINS	KNOWN NC FUNCTIONS	MGI NO.
Other (*cont'd*)	Myb	Gg (c-Myb)	Premigratory NC, migratory trunk NCCs,	NC migration, formation	97249
	Mybl2	Gg, Xl (B-Myb)	NCCs, PNS	Unknown	101785
	Myocd	Mm	Cardiac NC, OFT, VSMCs, aortic arch arteries	Control of vascular smooth muscle gene transcription	2137495
	Mkl1	Mm (MRTF-A)	Cardiac NC, OFT, VSMCs, aortic arch arteries	Control of vascular smooth muscle gene transcription	2384495
	Mkl2	Mm (MRTF-B)	Cardiac NC, OFT, VSMCs, aortic arch arteries	Control of vascular smooth muscle gene transcription	3050795
	Nfatc3	Mm	Schwann cells	Krox20 activation, Schwann cell differentiation	103296
	Nfatc4	Mm	Schwann cells	Krox20 activation, Schwann cell differentiation	1920431
	Nfyb	Dr	Cranial NCCs, pharyngeal arches, craniofacial cartilage	Unknown	97317

Other (*cont'd*)	Nr2f1	Mm (COUP-TFI)	PNS, cranial ganglia	Neuronal morphogenesis	1352451
	Nr2f2	Mm (COUP-TFII)	NC, PNS	Unknown	1352452
	Nr5a1	Mm (Sf1)	Adrenal medulla	Unknown	1346833
	Nr5a2	Mm (Ftf)	NCCs, pharyngeal arches	Unknown	1346834
	Rela (NFκB)	Xl, Rn	DRGs	Regulation of apoptosis, upregulation of Slug and Snail	103290
	Srf	Mm, Xl	Cardiac NC, VSMCs	Control of vascular smooth muscle gene transcription	106658
	Sp1	Mm, Hh, Gg	NC	Gata interaction	98372
	Sp3	Mm, Hh, Gg	NC	Unknown	1277166
	Sp4	Mm, Hh, Gg	NC	Unknown	107595
	Stat3	Mm, Xl	NCSCs	Unknown	103038
	Tcof1	Mm, Hs	Craniofacial skeleton	Craniofacial	892003
	Tead2	Mm	Prospective NC, VSMCs	Pax3 regulation	104904
	Tsc22d1	Mm	Pharyngeal arches, craniofacial skeleton, cranial ganglia, sympathetic ganglia, DRGs	Unknown	109127

Mm, Mus musculus; Hs, *Homo sapiens*; Rn, *Rattus norvegicus*; Gg, *Gallus gallus*; Dr, *Danio rerio*; Xl, *Xenopus Laevis*; Bp, *Bos primigenius*.

TABLE 1.2: List of abbreviations.	
ANS: autonomic nervous system	MPNST: malignant peripheral nerve sheath tumor
ARA: Axenfeld–Rieger anomaly	mRNA: messenger RNA
ARID: AT-rich interaction domain	NA: noradrenaline
ASD: anterior segment dysgenesis	NB: neuroblastoma
AV: atrioventricular	NC: neural crest
BAC: bacterial artificial chromosome	NCC: neural crest cell
bcNCSC: boundary cap-derived NCSC	NCSC: neural crest stem cell
bHLH: basic helix–loop–helix	NGF: neural growth factor
BMP: bone morphogenetic protein	NPB: neural plate border
C2H2: 2 cysteine, 2 histidine	NT: neural tube
cAMP: cyclic adenosine monophosphate	NTD: neural tube defect
CCHS: congenital central hypoventilation syndrome	OFT: outflow tract
CHD: congenital heart defect	PC12: pheochromocytoma cell line 12
ChIP: chromatin immunoprecipitation	PDL: periodontal ligament
CMT1: Charcot–Marie–Tooth neuropathy type 1	PI3K: phosphoinositide 3-kinase
CNS: central nervous system	PKA: protein kinase A
CRE: cAMP response element	pNCSC: palate-derived NCSC
DA: ductus arteriosus	PNS: peripheral nervous system
DHEA: dehydroepiandrosterone	DRG: dorsal root ganglia
DNA: deoxyribonucleic acid	E: embryonic day
MO: morpholino oligonucleotide	ECM: extracellular matrix

TABLE 1.2: (*continued*)	
MONC-1: mouse neural crest-1 (NCSC line)	EFT: Ewing family tumor
EMT: epithelial-to-mesenchymal transition	RNA: ribonucleic acid
ENS: enteric nervous system	RNAi: RNA inhibition
EPI-NCSC: epidermal-derived NCSC	ROSA26: reverse orientation splice acceptor 26
ER: estrogen receptor	ROSAMER: ROSA Myc-ER^T
ESC: embryonic stem cell	RPE: retinal pigmented epithelium
FGF: fibroblast growth factor	SA: sympathoadrenal
GFP: green fluorescent protein	SC: Schwann cell
GI: gastrointestinal	SHF: second heart field
HDAC: histone deacetylase	SKP: skin-derived precursor
HGF: hepatocyte growth factor	SMC: smooth muscle cell
HLH: helix–loop–helix	SMG: submandibular gland
HLHS: hypoplastic left heart syndrome	SNAG: Snail/growth factor independence-1
HMG: high mobility group	SNS: sympathetic nervous system
HSCR: Hisrchsprung's syndrome	IGF: insulin growth factor
pPNET: primitive peripheral neuroectodermal tumor	IP: immunoprecipitation
PSA: polysialic acid	iPSC: induced pluripotent stem cell
PTA: persistent truncus arteriosus	JNK: Jun N-terminal kinase
r: rhombomere	L-DOPA: L-dihydroxyphenylalanine
R26R: ROSA26 reporter	MHC: major histocompatibility complex
RA: retinoic acid	STAT: signal transducer and activator of transcription

TABLE 1.2: (*continued*)	
RARE: retinoic acid response element	TGF: transforming growth factor
VIP: vasoactive intestinal polypeptide	WS: Waardenburg syndrome
VPA: valproic acid	WS2: Waardenburg syndrome type 2
VSMC: vascular smooth muscle cell	WS4: Waardenburg syndrome type 4

• • • •

CHAPTER 2

AP Genes

2.1 AP-1

Activator protein 1 (AP-1) is a heterodimer typically consisting of the oncogenic transcription factors Jun (often called c-Jun) and Fos (c-Fos) but can also include other less common AP-1 subfamily members. AP-1 binds DNA at conserved response element sequences via a leucine zipper domain (Abate and Curran, 1990). Both Jun and Fos are expressed in some NC lineages, and AP-1 does have activity in several NC lineages. In *Xenopus*, AP-1 induces transcription of Zic3, a regulator of neural and NC development whose promoter contains several AP-1 binding sites. This activity requires canonical AP-1; other homo- or heterodimeric combination of proteins, such as Jun/Jun, JunD/FosB, or JunD/Fra-1 are unable to induce Zic3. Zic3 expression induced by Activin is blocked by inhibiting AP-1 activity with knockdown of Jun, and this induction is rescued by coinjection with *Jun* mRNA (Lee et al., 2004a).

Although little is known regarding AP-1 in NC development, several studies indicate an important role for Jun in NCCs. RA treatment of mouse embryos results in defects similar to avian NC ablation, resulting in suppressed NC migration and proliferation both in vivo and in vitro, particularly with regard to the cardiac NC. In NCCs, Jun N-terminal kinase (JNK) activation is severely repressed by RA. Inhibition of upstream activation of JNK or Jun downstream responses reduces NCC outgrowth. JNK signaling pathway and Jun activation are critical for cardiac NC outgrowth and are potential targets for the action of RA (Li et al., 2001). In mice, a role for Jun in the cardiac NC is suggested by analysis of mutant embryos. *Jun*-null embryos display a persistent truncus arteriosus, a common cardiac NC defect. These embryos die around E13.0, likely due to a combination of heart and liver defects, with extensive apoptosis of hematopoietic cells and hepatoblasts, suggesting an additional role for Jun in hepatocyte turnover and regulation of cell survival (Eferl et al., 1999). Consistent with a role in the cardiac NC, Jun is expressed in NC-derived vascular smooth muscle cells (VSMCs) of the aortic arch arteries. After surgical ablation of NC in the avian embryo, mesoderm-derived VMSCs replace NC-derived VMSCs populating the OFT. When comparing these two VMSC populations, either grown in culture under identical conditions or freshly removed from an embryonic vessel, the surrogate mesoderm-derived VSMCs were found to express up to 15 times more Jun (Gadson et al., 1993). Finally, AP-1 elements have been shown to contribute to induction of the *SM22α* promoter by TGFβ in the Monc-1 NCC line (Chen et al., 2003).

Jun is also expressed in NC-derived Schwann cells. During development, c-Jun is required for Schwann cell proliferation and regulation of Schwann cell survival. Upon initiation of the myelination program, Jun is downregulated by Krox-20. Forced expression of Jun in Schwann cells prevents myelination. After peripheral nerve injury, Schwann cells demyelinate, proliferate, and dedifferentiate, assuming a phenotype similar to that of immature Schwann cells. Jun is required for dedifferentiation of Schwann cells toward a more immature Schwann cell state and is also necessary for neuron regeneration (Mirsky et al., 2008). Several cell lines have been used to determine the involvement of Jun, and to a lesser extent, Fos, in NC lineages. The NC-derived rat pheochromocytoma cell line PC12 expresses the nerve growth factor receptors Ntrk1 and p75. Upon NGF treatment, Ntrk1 is phosphorylated on tyrosine residues and the early responsive genes *Egr1*, *Nr4a1*, *Fos*, and *Jun* are induced (Tazi et al., 1995). NC-derived Schwann cells express only the low-affinity receptor p75, and Egr1, Nr4a1, and Fos are activated upon NGF treatment (Matheny et al., 1992). However, this effect is not seen in the NC-derived rat schwannoma cell line, JS1, which expresses NGFI-A and NGFI-B constitutively and does not express Fos, probably due to an activated transcriptional repressor specific to this cell line (Matheny et al., 1992). In the NC-derived neuroendocrine chromaffin cells of the bovine adrenal medulla, expression of the neuropeptide Enkephalin is regulated in part by specific binding of AP-1 family members, including c-Jun, JunD, and possibly a Fos protein (MacArthur, 1996). Fos expression, in addition to other protooncogenes, also is augmented during the differentiation of NC-derived human neuroblastoma cells (Thiele et al., 1988).

2.2 *AP-2α (Tcfap2a)*

There are five identified *AP-2* genes (*Tcfap2a-e*, also known as AP-2α, β, γ, δ, and ε), and these genes share a high degree of similarity. However, before the existence of multiple AP-2 genes was recognized, most early NC studies were presumably analyzing the combined expression or effects of all three isoforms that are present in the NC (Tcfap2a, b, and c), or perhaps only Tcfap2a, the predominant AP-2 isoform, alone. In this section, where the designation AP-2 is used, it should be noted that this likely refers to AP-2α but will occasionally refer to all AP-2 isoforms (Figure 2.1).

2.2.1 Mouse *AP-2α*

During NT closure in the mouse embryo, from E8.5 to E12.5, AP-2 is expressed in ectoderm and cranial NCCs, including cells that will become the facial mesenchyme and sensory ganglia of the head and spinal cord (Kohlbecker et al., 2002; Mitchell et al., 1991; Morriss-Kay, 1996; Talbot et al., 1999; Williams and Tjian, 1991). *AP-2*-null mice die perinatally with cranial closure failure and severe dysmorphogenesis of the face, skull, sensory organs, OFT of the heart, and cranial ganglia. Failure of cranial closure between E9 and E9.5 coincided with increased apoptosis in the midbrain, anterior hindbrain, and proximal mesenchyme of the first pharyngeal arch (Brewer et al., 2002c;

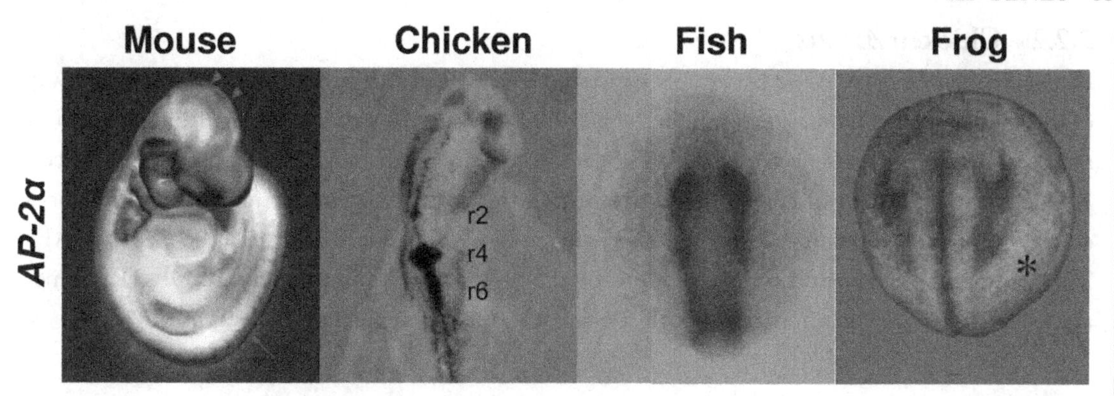

FIGURE 2.1: Whole-mount in situ hybridizations show *AP-2α* in mouse, chicken, fish, and frog embryos. Mouse: Solloway and Robertson, 1999, 9.5-dpc embryo, lateral view; chicken: Theveneau et al., 2007, 13-somite-stage embryo, dorsal view, anterior at top (all subsequent dorsal views will be oriented with anterior at top); zebrafish: Eroglu et al., 2006, 13-somite embryo, dorsal view; *Xenopus*: Schuff et al., 2007, stage 23 embryo, dorsal view.

Kohlbecker et al., 2002; Morriss-Kay, 1996; Schorle et al., 1996; Zhang et al., 1996). A subset of *AP-2α* heterozygous animals also develop midbrain exencephaly and have reduced skull size and malformed cranial vault bones, possibly due to a downregulation of the remaining allele suggesting the involvement of an upstream modifier gene or sensitivity to gene dosage (Kohlbecker et al., 2002). Analysis of an *AP-2α-IRES-lacZ* knock-in allele reveals expression in the cardiac NC population between E9.5 and 10.5 (in addition to other NC derivatives), and this expression was mostly extinguished by E11.5 at which time the cardiac NC has migrated into the OFT, suggesting that AP-2α functions in the cardiac NC before E11.5. The distribution of AP-2α-expressing cardiac NC appeared identical in normal and mutant embryos, further suggesting that AP-2α is not required for NCC migration or early survival. An open hypothesis is that AP-2α is involved in an interaction between NC and surrounding tissues in the subpharyngeal region, thereby promoting normal OFT morphogenesis (Brewer et al., 2002c). An NC-specific deletion of a floxed *AP-2α* allele using *Wnt1-Cre* also results in frequent perinatal lethality associated with NT closure defects and cleft secondary palate. A small fraction of these mutant mice survive into adulthood but have retarded craniofacial growth, abnormal middle ear development, and defects in pigmentation. However, several NC defects affecting the head and heart that might be predicted based on the *AP-2α*-null embryonic phenotype were not observed, suggesting functional redundancy with other *AP-2* genes in these derivatives (Brewer et al., 2004). AP-2α is also expressed and upregulated in parts of the hair follicle during its early morphogenesis (Panteleyev et al., 2003), and adult stem cell populations have been isolated from cells of NC origin residing in the hair follicle (Sieber-Blum and Hu, 2008; Sieber-Blum et al., 2006).

2.2.2 Chicken *AP-2α*

In the chicken embryo, AP-2 is strongly expressed in the head throughout the migration pathway of the cranial NCCs. Later, expression in the facial mesenchyme is strongest in the frontonasal mass and lateral nasal prominences and downregulated in the maxillary and mandibular prominences. Retinoic acid treatment inhibits outgrowth of the frontonasal mass and lateral nasal prominences concurrently with a downregulation of AP-2 and an increase in apoptosis around the nasal pit. AP-2 is involved in outgrowth and transcription could be regulated by factors such as FGFs (Payson et al., 1998) present in the ectoderm of both the face and the limb (Shen et al., 1997).

2.2.3 Zebrafish *AP-2α* Mutants

Zebrafish homozygous for the *AP-2α^{lockjaw}* loss-of-function mutation have defects in all NC derivatives, including craniofacial cartilages and pigment cells (Knight et al., 2003; Knight et al., 2004). NC specification and migration are disrupted in *AP-2α^{lockjaw}* mutant embryos, and there is a transient period of apoptosis in NCCs before and during migration (Knight et al., 2003). In cranial NC derivatives, gene expression in the mandibular arch is unaffected, but there are severe reductions in Dlx2, Hoxa2, and Hoxb2 expression in the hyoid arch, reflecting reversion to a mandibular fate (Knight et al., 2004; Knight et al., 2003). The cranial NC defects are cell autonomous but cause disruptions in surrounding non-NC mesoderm. Independent of Hox factors, *AP-2α^{lockjaw}* mutants also have defects in a subset of pigment cells (the melanophores and iridophores, but not the xanthophores). Early melanophores do not develop due to melanophore specification defects and subsequent migration defects concomitant with loss of Kit function (Knight et al., 2004). Like the *AP-2^{lockjaw}* mutation, the *AP-2^{montblanc}* mutation (Barrallo-Gimeno et al., 2004) affects all NC derivatives. Initial NC induction, specification, and early migration appear undisturbed, but the craniofacial NC derivatives in pharyngeal arches 2–7 later fail to express a normal NC gene program consisting of *Sox9a*, *Wnt5a*, *Dlx2*, *Hoxa2*, and *Hoxb2*. NCCs migrate normally into the first pharyngeal arch while preotic and postotic NCCs start to migrate but fail to descend to the pharyngeal region of the head, and the number of neuronal and glial NC derivatives is greatly reduced (Barrallo-Gimeno et al., 2004). In the absence of AP-2α, a subset of NCCs is unable to undergo differentiation; these cells undergo apoptosis. Surviving NCCs in *AP-2^{montblanc}* mutant embryos proliferate normally and later differentiate to individual derivatives (Barrallo-Gimeno et al., 2004; Holzschuh et al., 2003). Together, analyses of these two mutants suggest that AP-2α is required for early steps in NC development and survival of a subset of NC derivatives (Knight et al., 2003) and is essential to activate the normal developmental program in pharyngeal arches 2–7 and in trunk NC (Barrallo-Gimeno et al., 2004). This interpretation is consistent with data from *AP-2α* morphants; those embryos undergo normal early morphogenesis and NC induction but have defects in NC derivatives including the jaw cartilage, enteric neurons, sympathetic neurons, and pigment cells (O'Brien et al., 2004). In AP-2α morphants, multiple steps in melanophore development are affected. Expression of c-Kit is reduced and embryonic melanophores are fewer with attenuated migration; but in contrast to c-Kit mutant embryos, the melanophores are small and underpigmented, suggesting differentiation is regulated by AP-2α through cell nonautonomous targets (O'Brien et al., 2004).

2.2.4 *Xenopus AP-2α*

Xenopus AP-2α mRNA is restricted to the prospective epidermis by the end of gastrulation, under the control of BMP signals, and elevated expression in the prospective NC begins at this time (Luo et al., 2002). The AP-2α expression domain is larger than for other NC factors (Luo et al., 2003). In vivo and in vitro studies demonstrate that AP-2α has a central role in NC induction. In intact *Xenopus* embryos and animal cap ectodermal explants, BMP and Wnt signaling positively regulate AP-2α, and this activation is required for NC induction (Luo et al., 2003). Like other NC regulatory factors, AP-2α induces NC in isolated ectoderm with active Wnt signaling and attenuated BMP signaling. Ectopic expression of AP-2α causes expression of epidermal genes in neuralized ectoderm (Luo et al., 2002) and is sufficient to activate strong expression of NC-specific genes such as *Slug* and *Sox9* in the presence of Wnt signaling (Luo et al., 2003). Knockdown of AP-2 function, or expression of a dominant-negative AP-2, leads to the reverse: loss of epidermal gene expression and gain of neural gene expression (Luo et al., 2002) causing a severe reduction in the NC territory (Luo et al., 2003). AP-2α expression, along with inactivation of NC inhibitory factors such as Dlx3, is thought to establish a feedback loop consisting of AP-2α, Sox9, and Slug necessary for initiation and maintenance of NC specification (Luo et al., 2003).

2.2.5 Regulation of AP-2

AP-2 expression in NC lineages is induced by retinoic acid (RA) across multiple vertebrate species (Chazaud et al., 1996; Holzschuh et al., 2003; Shen et al., 1997; Williams and Tjian, 1991; Zhang et al., 1996). As described above from *Xenopus* studies, AP-2α, β, and γ are also regulated by Wnt/β-catenin and BMP signals (Luo et al., 2003; Zhang et al., 2006b). In *Xenopus*, Klf12 (previously referred to as AP-2 repressor or AP-2rep) restricts AP-2 expression and is expressed in all neural tissue at the neurula stage, whereas AP-2 is predominantly expressed in cranial NC. Ectopic injection of *Klf12/AP-2rep* RNA reduces endogenous expression of AP-2 in the NC, suggesting that differentiation of neural tissues requires repression of AP-2 (Gotoh et al., 2003). More recent studies demonstrated that Klf12/AP-2rep is also involved in mesodermal morphogenesis at the gastrula stage, operating through Brachyury and the Wnt pathway (Saito et al., 2009). Other potential upstream regulators of AP-2 include Hand2 and members of the Stat family. An intronic enhancer in the mouse and human *AP-2α* locus directs expression of a LacZ reporter to the facial processes and the distal limb mesenchyme; this enhancer contains a conserved Stat binding site. This site is necessary for expression of the transgene within the NC-derived facial prominences but not in the limbs (Donner and Williams, 2006). Finally, expression of AP-2α is significantly reduced in zebrafish *Hand2*$^{hands\ off}$ mutant embryos, suggesting a role for AP-2α as a downstream effector during noradrenergic differentiation of sympathetic neurons (Lucas et al., 2006).

2.2.6 AP-2 Targets

In mammalian cells, AP-2 protein binds the palindromic AP-2 consensus sequence as a dimer, but this binding is not necessary for dimerization. AP-2 most commonly acts as a transcriptional activator, and this activity is mediated by an amino-terminal proline-rich domain (Williams and Tjian, 1991). AP-2 most commonly regulates neural and epithelial gene transcription (Zhang et al., 1996). One of the partners of AP-2 in gene regulation is Cited2. Cited2 interacts with and coactivates all isoforms of AP-2 (Bamforth et al., 2001; Braganca et al., 2002) and is required for NC and NT development. *Cited2*-null embryos die with a phenotype very similar to AP-2α-null embryos, including exencephaly, abnormal cranial ganglia, cardiac malformations, and adrenal defects, and an increase in apoptosis in the midbrain region. These embryos also have a reduction in NCC expression of the AP-2 target gene *Erbb3*, further suggesting that *Cited2*-null defects are due at least in part to its role as an AP-2 coactivator (Bamforth et al., 2001; Weninger et al., 2005). Further supporting this hypothesis, transactivation by AP-2 isoforms is defective in *Cited2*-null embryonic fibroblasts and can be rescued by ectopically expressed Cited2 (Bamforth et al., 2001). In addition to Cited2, Cited4, also a transactivator, physically interacts with all AP-2 isoforms in vitro and strongly coactivates all AP-2 isoforms in Hep3B cells. In some cell lines, Cited4 is significantly weaker than Cited2 for AP-2γ co-activation, suggesting that different cell type- and AP-2 isoform-specific combinations with the coactivators Cited2 or Cited4 may be used for differential modulation of AP-2 function in different tissues (Braganca et al., 2002).

There are several identified targets of AP-2 including *Erbb3* and *Hoxa2* (Maconochie et al., 1999; Weninger et al., 2005). NC-specific enhancer elements, including one element with an AP-2 binding site, were identified for Hoxa2, which is expressed in a subset of cranial NCCs. Mutation or deletion of the AP-2 binding site in the *Hoxa2* enhancer abrogates reporter expression in the cranial NCCs but not in the hindbrain where AP-2 is not expressed. In cell culture and in transgenic embryos, AP-2 family members are able to transactivate reporter expression. Reporter expression is not abolished in *AP-2α*-null embryos, suggesting redundancy with other AP-2 family members for activation of the *Hoxa2* enhancer (Maconochie et al., 1999). AP-2 is expressed and functions in the adrenal medulla, where epinephrine is produced. AP-2 stimulates expression of *phenylethanolamine N-methyltransferase (Pnmt)*, the gene encoding the biosynthetic enzyme for epinephrine. Stimulation of Pnmt by AP-2 requires glucocorticoids and may be mediated through interaction of AP-2 with activated type II glucocorticoid receptors. Mutation of AP-2 and/or glucocorticoid receptor binding elements within the *Pnmt* promoter disrupts the ability of AP-2 and glucocorticoids to induce *Pnmt* promoter activity (Ebert et al., 1998). Although most target genes are activated by AP-2, there is evidence that AP-2 can also act as a transcriptional repressor; examination of genes that are prematurely expressed in *AP-2α* mutant mice led to the identification of a set of genes repressed by AP-2α. Most of these inhibit proliferation and induce differentiation and apoptosis. One of the genes isolated, *Klf4*, is implicated in induction of terminal differentiation and growth regulation (Pfisterer et al., 2002). Using the *Xenopus* animal cap assay, misexpression of a hormone inducible AP-2α fusion protein identified downstream targets of this factor. Most of these putative targets were expressed at high levels in the NC but were also expressed in the epidermis and in other tissues in which AP-2α expression has not been detected, suggesting modular control involving both AP-2-dependent and AP-2-independent components. This work also identified a set of genes expressed in the neural plate that were repressed by AP-2α (Luo et al., 2005). Inca (induced in neural crest by AP-2) is one of the upregulated genes identified in this microarray screen. Inca is expressed primarily in premigratory and migratory NCCs in *Xenopus*, zebrafish, and mouse embryos. Knockdown of Inca expression in *Xenopus* and zebrafish revealed a requirement for Inca in a subset of NCCs forming craniofacial cartilage. Inca interacts directly with the NC-expressed p21-activated kinase 7 (Pak7) and may cooperate in regulation of cytoskeletal organization and cell adhesion in craniofacial NCCs (Luo et al., 2007).

2.2.7 AP-2 Binding Sites

In addition to a number of tested and verified AP-2 target genes, many others are suggested based on the presence of AP-2 consensus binding sequences and shared expression domains. These include *Mcam*, a cell surface glycoprotein associated with human malignant melanoma (Sers et al., 1993), *proto-Ret*, a protooncogene highly expressed in tumors of NC origin (Itoh et al., 1992), and finally, *Sox10*, which has AP-2 binding sites in enhancers mediating Sox10 expression in the otic vesicle, oligodendrocytes, and several NC derivatives including the developing PNS and the adrenal gland (Werner et al., 2007).

2.2.8 AP-2 as an NC Marker

Because of its critical early role in the NC, expression of AP-2α is often used as a marker for premigratory and migratory NCCs in many vertebrate species (Brewer et al., 2004; Conway et al., 2000; Epperlein et al., 2000; Gajavelli et al., 2004; Ishii et al., 2005; Lawson et al., 2000; Minarcik and

Golden, 2003; Mori-Akiyama et al., 2003; Saint-Jeannet et al., 1997). It is used as an NC marker in mouse *Pax3*^{Splotch2H} mutant embryos to track reductions in migratory NCCs (Conway et al., 2000), in *Xenopus* to monitor induction of NC by Wnt1, Wnt3a, Noggin, and Chordin (Saint-Jeannet et al., 1997), and in mammalian cell culture, it is one of the factors used to assess NC fate during directed differentiation of rat cortical NSCs toward the NC lineage (Gajavelli et al., 2004). AP-2α is expressed in human embryos and a wide variety of neuroectodermal tumors, and increased expression may correlate with the extent of differentiation in these NC tumors. Expression of the AP-2α target Erbb3 also correlated with AP-2α expression in neuroectodermal tumors (Gershon et al., 2005; Schulte et al., 2008).

2.3 *AP-2β (Tcfap2b)*

Murine AP-2β is closely related to AP-2α (76% homology). Expression of both genes starts at E8 in lateral head mesenchyme and extraembryonic trophoblast. Starting at E11, expression of AP-2β, together with AP-2α, is detected in the skin, facial mesenchyme, spinal cord, cerebellum, and renal tubular epithelia. Expression of AP-2β alone occurs in the midbrain, sympathetic ganglia, adrenal medulla, and cornea (Moser et al., 1997). Human AP-2β is also expressed in NCC lineages and neuroectodermal cells. Within the monoaminergic systems, several genes have binding sites for AP-2β. AP-2β is associated with serotonergic phenotypes in the rat (Damberg et al., 2005). In zebrafish, AP-2α (tfap2a) and AP-2β (tfap2b) are essential in development of the facial ectoderm and for signals from the epithelium that induce skeletogenesis in NCCs. Zebrafish embryos deficient for both AP-2α (tfap2a) and AP-2β (tfap2b) have defects in epidermal cell survival and a loss of NCC-derived cartilage (Knight et al., 2005). Cartilage defects arise after NCC migration, during skeletal differentiation, and can be rescued by transplantation of wild-type ectoderm. These data suggest that AP-2 proteins may play two distinct roles in cranial NCCs: an early cell-autonomous function in specification and survival of NCCs, and later nonautonomous function regulating ectodermal signals that induce NCCs to undergo skeletogenesis (Knight et al., 2005). In *Xenopus*, AP-2β is expressed in NC at late gastrula to early neurula stages and is essentially NC-specific. AP-2β is induced by Wnt and attenuated BMP signals and binds a consensus AP-2 recognition site from an epidermal keratin gene (Zhang et al., 2006b).

Only AP-2β is predominantly expressed in the sympathetic ganglia of developing mouse embryos, supporting its role in sympathetic development. *AP-2β*-null mice expressed significantly reduced levels of both norepinephrine and dopamine β-hydroxylase (Dbh) in the PNS. In utero administration of an NA intermediate during pregnancy significantly rescues the neonatal lethality of *AP-2β*-null mice, suggesting that NA deficiency is the main cause of lethality of the *AP-2β*-null mice. Forced expression of AP-2β in NCSCs directs their differentiation toward an NA neuron fate (Hong et al., 2008c). In addition to its role in neuronal development, AP-2β plays an important role in heart development. Char syndrome, a human syndromic form of patent ductus arteriosus (DA) with facial dysmorphology and fifth-finger clinodactyly, with occasional occipital bone defects and parasomnia, is caused by mutation of human *AP-2β* (Mani et al., 2005). AP-2β is uniquely expressed in the mostly NC-derived VSMCs of the DA in mouse. Edn1 and Hif2α are also highly enriched in smooth muscle of the DA at E13.5 and are dependent on AP-2β. Hif2α is a negative regulator of AP-2β-induced transcription, disrupting protein–DNA interactions, suggesting a negative feedback loop (Ivey et al., 2008). There is a single reported case of a pigmented peripheral nerve sheath tumor with a high degree of melanocytic differentiation, likely a tumor of NC origin, and this tumor expressed AP-2β and c-Ret (Dammer et al., 1997).

2.4 *AP-2γ (Tcfap2c)*

AP-2γ was identified as a novel early RA-induced gene in the mouse embryonal carcinoma cell line P19. It is closely related to AP-2α and can activate transcription using an AP-2 binding site. *AP-2γ* transcripts are detected in the NPB and in extraembryonic tissues, and by E8.0–E8.5, AP-2γ is expressed in premigratory and migrating NCCs, and later restricted to the facial mesenchyme, especially around the embryonic mouth and nasal cavities (Chazaud et al., 1996). *AP-2γ* heterozygous mice are viable and fertile but are smaller than normal at birth. *AP-2γ* homozygous null embryos die between E7 and E9 likely due to a trophectoderm defect (Werling and Schorle, 2002). In *Xenopus*, AP-2γ is expressed in outer epidermal cells and the prospective NC region at late gastrula to early neurula stages and is induced by Wnt and attenuated BMP signals (Zhang et al., 2006b). In zebrafish, AP-2γ (tfap2c) is expressed in nonneural ectoderm including transient expression in NC. Knockdown of AP-2γ does not perturb development, but simultaneous inhibition of AP-2α and AP-2γ prevents NC induction, supporting a conserved role for AP-2 activity in NC induction (Hoffman et al., 2007; Li and Cornell, 2007).

2.5 *AP-2δ (Tcfap2d) AND AP-2ε (Tcfap2e)*

Unlike AP-2α, β, and γ, AP-2δ and AP-2ε have not been detected in the NC or its derivatives.

• • • •

CHAPTER 3
bHLH Proteins

Basic helix–loop–helix (bHLH) genes are transcription factors that typically bind to a consensus E box sequence. These factors often form homodimers or heterodimers with other bHLH factors, allowing for a broader degree of combinatorial tissue-specific gene regulation.

3.1 Ascl1

Ascl1 (achaete-scute-complex homolog 1; also known as Mash1, Hash1, or Cash1) is the vertebrate homolog of achaete-scute, an important regulator of *Drosophila* neuronal development (Dambly-Chaudiere and Vervoort, 1998). Mouse and rat Ascl1 is expressed early in the development of the ANS in precursors of sympathetic, parasympathetic, and enteric neurons (Anderson, 1994; Lo et al., 1991; Lo et al., 1994), before neuronal differentiation, and then extinguished after differentiation (Lo et al., 1991; Shoba et al., 2002). In humans, Ascl1 is detected in sympathetic cells only at the earliest embryonic ages examined, likely due to loss of expression after neuronal differentiation (Gestblom et al., 1999). Ascl1 is required for specification and differentiation of ANS neurons as part of a regulatory network including Phox2a, Phox2b, and Hand2 (also in this network are Tlx2, Tlx3, and Gata2/3) (Bachetti et al., 2005; Howard, 2005; Lo et al., 2002; Morikawa et al., 2007; Perez et al., 1999; Pogoda et al., 2006; Tsarovina et al., 2004). *Ascl1*-null mice die at birth with breathing and feeding defects. Sympathetic ganglia, parasympathetic ganglia, and enteric ganglia of the foregut are severely reduced if not gone, and enteric ganglia of the stomach and hindgut are partially affected (neuronal differentiation in the ENS occurs after NCCs localize to the ANS anlagen), due to arrested development and failure of neuronal precursors to differentiate (Anderson, 1993; Anderson, 1994; Guillemot et al., 1993; Lo et al., 1994). In contrast, differentiation of autonomic glia is normal (Lo et al., 1994). In cultures of NT-derived, primitive PNS progenitors, Ascl1 promotes autonomic marker expression (Lo et al., 2002). Ascl1 alone does not commit multipotent NCCs to a neuronal fate but instead promotes differentiation of already-committed neuronal progenitors (Sommer et al., 1995).

3.1.1 Ascl1 in the Sympathoadrenal Lineage

Ascl1 also marks the developing sympathoadrenal NC lineage, including thyroid C cells (NC-derived neuroendocrine cells with neuronal and serotonergic characteristics) and adrenal medulla chromaffin cells. Thyroid C cells and serotonergic enteric neurons may arise from a common Ascl1-expressing sympathoadrenal progenitor, and Ascl1, in addition to being required for differentiation of serotonergic enteric neurons, is needed to establish the C-cell phenotype; *Ascl1*-null mice have a greatly reduced number of C cells (Lanigan et al., 1998). Cultured rat C cells initially express Ascl1, but it quickly becomes undetectable after the C cells are placed in culture, consistent with observed neuronal differentiation (Clark et al., 1995). Murine Ascl1 is required for general chromaffin cell differentiation and catecholaminergic differentiation (Huber et al., 2002a; Huber et al., 2002b). During early development, the number of adrenal medullary Phox2b-positive progenitor cells is initially unaffected in *Ascl1*-null mice, but by birth, mutants only have one third the amount present in wild type, and tyrosine hydroxylase (Th)-expressing (catecholaminergic) cells are greatly reduced. Most cells in the adrenal medulla of Ascl1-null mice do not contain chromaffin granules and have an immature, neuroblast-like phenotype, with very few ever expressing the distinguishing chromaffin cell gene, *Pnmt* (Huber et al., 2002a; Huber et al., 2002b; Huber et al., 2005). Both Ascl1 and Phox2b are expressed by sympathoadrenal progenitors, but these two proteins are required independently and have distinct requirements for development of chromaffin cells and sympathetic neurons (Huber et al., 2005; Unsicker et al., 2005).

3.1.2 Regulation of Ascl1 by BMP, cAMP, and Notch Signaling

Ascl1, with Hand2, Phox2a, Phox2b, and Gata2/Gata3, is induced in NC-derived neuronal progenitors by BMPs (secreted from tissues to which autonomic neuron progenitors migrate) during development of sympathetic neurons (Anderson et al., 1997; Lo et al., 1998; Lucas et al., 2006; Pisano et al., 2000; Schneider et al., 1999). In cultured NCSCs, Bmp2 and Bmp4 induce Ascl1 and promote autonomic neuronal differentiation (Greenwood et al., 1999; Lo et al., 1997; Lo et al., 1998; McPherson et al., 2000). Expression of Ascl1 in postmigratory NCCs allows nonneuronal cells to remain competent for neuronal differentiation in the presence of Bmp2, and Bmp maintains Ascl1 expression (Lo et al., 1997). Constitutive Ascl1 expression maintains competence of NC progenitors to respond to later Bmp2 signals and undergo neuronal differentiation (Lo et al., 1997). Bmp2 promotes development of cultured primary NCCs toward the sympathoadrenal lineage, including induction of Ascl1, Phox2a, Phox2b, and the catecholaminergic markers Th and catecholamines (Bilodeau et al., 2001). Bmp2, Bmp4, Bmp7, and elevated cAMP synergistically stimulate development of the sympathoadrenal lineage in NCCs. Together with Bmp2, moderate activation of cAMP signaling promotes sympathoadrenal cell development and expression of *Phox2a*, a sympathoadrenal lineage-determining gene. In contrast, strong activation of cAMP signaling represses sympathoadrenal cell development and Ascl1, Phox2a, and Phox2b (Bilodeau et al., 2000). In murine NCSCs, Ascl1 is more sensitive to Notch-mediated inhibition of neurogenesis and cell cycle arrest than the sensory lineage neuronal genes *Neurogenin1* (*Neurog1*) and *Neurog2*, suggesting a differential role for these genes regulated by Notch signaling (Lo et al., 2002). Ascl1 is suppressed by Notch signaling, a repression necessary to maintain ENS progenitors by restricting NC neuronal differentiation. The number of Ascl1-expressing enteric NCCs is increased in *Protein O-fucosyl transferase (Pofut)*-null embryos lacking this enzyme involved in Notch cleavage (Okamura and Saga, 2008).

3.1.3 Ascl1 Interaction with Other Transcription Factors and Targets

There is evidence of considerable interplay between factors of the regulatory network defining/determining autonomic neurogenesis that includes Ascl1, Phox2a, Phox2b, Hand2, and Gata2/3. In the ENS and sympathetic ganglia, Phox2b is needed to maintain Ascl1 expression (Pattyn et al., 1999), and Phox2b upregulates expression of Ascl1 in combination with Nkx2.2 (Dubreuil et al., 2002). Although Phox2a and Ret are induced by Bmp2, constitutive Ascl1 in NCSCs can induce these factors without Bmp2 induction. Phox2a in autonomic ganglia is strongly reduced in *Ascl1*-null embryos (Lo et al., 1998). The *Phox2a* promoter has an E-box element that can be bound by Ascl1 (which, as described above, is induced by BMPs) near elements activated by components of the cAMP signaling pathway, suggesting a possible mechanism for synergy between BMP and cAMP signaling (Benjanirut et al., 2006). Sox10 is required in vivo for induction of Ascl1 and Phox2b (Kim et al., 2003), but Ascl1 is a strong repressor of Sox10, indicating a negative feedback loop that allows Ascl1-expressing cells to continue neuronal differentiation (Okamura and Saga, 2008). *In vitro*, Ascl1 activates the *Ret* promoter, important for intestinal innervation (Bachetti et al., 2005). The noradrenergic marker genes *Th* and *Dbh* are also downstream of Ascl1 (Schneider et al., 1999).

3.1.4 Ascl1 in Human Disease

Congenital central hypoventilation syndrome (CCHS) is often associated with other ANS dysfunction, suggesting involvement of genes expressed broadly in the ANS, such as those in the Ascl1-Phox-Ret pathway. A rare heterozygous nucleotide substitution in Ascl1 was found in three patients with CCHS. Reproduction of these mutant Ascl1 alleles in an in vitro model of NCC noradrenergic differentiation disrupted noradrenergic neurogenesis (de Pontual et al., 2003). Human neuroblastomas, sympathetic NC-derived tumors, often express Ascl1 (about two thirds of all neuroblastomas) (Arvidsson et al., 2005; Gestblom et al., 1999; Ichimiya et al., 2001). In differentiating neuroblastoma cells, Ascl1 is downregulated with a corresponding upregulation of the transcriptional repressor Hes1, known to bind the *Ascl1* promoter (Jogi et al., 2002). *Ascl1* transcripts are also downregulated in several neuroblastoma lines before RA-induced differentiation, but neuroblastoma cell lines without endogenous Ascl1 do not respond to RA, suggesting the downregulation of

Ascl1 may be an important step in promoting neuronal differentiation of neuroblastomas (Ichimiya et al., 2001). Ascl1 is also expressed in pheochromocytomas, NC-derived adrenal medullary thyroid tumors, and small cell lung cancers, all of which have characteristics of both neuronal and endocrine cells (Ball et al., 1993; Cairns et al., 1997).

3.2 Hand1

The Hand (heart and neural crest derivatives expressed) transcription factor group is made up of Hand1 and Hand2, both of which are bHLH factors expressed within the NC and derivatives. A modified yeast two-hybrid screen of a mouse embryonic cDNA library for proteins binding the *Drosophila* bHLH E protein daughterless/Tcf3 identified Hand1 (previously known as eHand for its early *e*xtraembryonic expression, to distinguish it from Hand2/dHand, which has early *d*eciduum expression) (Hollenberg et al., 1995). Mouse Hand1 is widely expressed in extraembryonic tissues throughout development, and in parts of the heart, autonomic ganglia, gut, and pharyngeal arches. *Hand1* mRNA can be detected in the embryonic heart at E8.5 until E10.5 when it declines abruptly except in the developing valves, where it persists until at least E13.5 (Cserjesi et al., 1995; Hollenberg et al., 1995). The strongest embryonic expression of Hand1 beyond the heart is in NC-derived tissues, especially first pharyngeal arch and derivatives, ANS, ENS, and sympathoadrenal lineage (Cserjesi et al., 1995; Hollenberg et al., 1995). In the adult, Hand1 expression has only been described in smooth muscle cells of the gut (Hollenberg et al., 1995). Both Hand1 and Hand2 are expressed in mesodermal and NC-derived structures (OFT and aortic arch arteries) of the embryonic heart. In the mouse, but not in the chicken embryo, Hand1 and Hand2 are expressed complementarily in the mesodermal heart tissue, restricted to the developing left and right ventricles, respectively (Srivastava et al., 1997). In *Xenopus*, Hand1 and Hand2 are coexpressed in the cardiac mesoderm, without restriction to a particular ventricle as seen in mice, and are also expressed in the lateral mesoderm without any left–right asymmetry. Within the pharyngeal arches, Hand2 is expressed in a broader domain than Hand1, similar to that in mouse (Angelo et al., 2000). In zebrafish, there is no Hand1 homolog (Angelo et al., 2000).

Mouse embryos homozygous for a *Hand1*-null lacZ knock-in allele (*Hand1*$^{lacZ/lacZ}$) die between E7.5 and E9.5 with yolk sac and trophoblast defects because of extraembryonic mesoderm deficiency. Although these embryos die before heart development is complete, early development of the heart is disrupted, with no progress beyond the cardiac-looping stage (Firulli et al., 1998; Riley et al., 1998). The lethal extraembryonic defects can be rescued by aggregation of mutant embryos with wild-type tetraploid embryos; the tetraploid cells contribute to the extraembryonic lineages, allowing for later assessment of heart development. Trophoblast-rescued *Hand1*-null embryos die by E10.5 due to cardiac failure including abnormal looping and differentiation of the ventricular myocardium (Riley et al., 1998). In vitro, *Hand1*-null ES cells differentiate into beating cardiomyocytes expressing cardiac myosin and other cardiac-specific transcripts, indicating that Hand1 is not essential for ventricular cardiomyocyte differentiation (Riley et al., 2000). In chimeras derived from *Hand1*-null ES cells injected into *Rosa26*^{lacZ} blastocysts wild type for Hand1, mutant cells were underrepresented in parts of the developing left ventricular myocardium and in the NCC-derived OFT and first pharyngeal arch, suggesting Hand1 is needed cell-autonomously in both mesodermal and NC derivatives (Riley et al., 2000). The phenotype of *Hand1*-null embryos (failure of heart looping, severely hypoplastic left ventricle, and OFT defects) is similar to human hypoplastic left heart syndrome (HLHS), and in most of the HLHS patients analyzed, a frameshift mutation in the Hand1 bHLH domain was identified. This mutation disrupted transcriptional modulation of a Hand1-binding reporter construct (Reamon-Buettner et al., 2008). Hand1 expression is reduced in hearts of Hif1α-deficient mice and may be the cause for the disruption of cardiac looping in those embryos (Compernolle et al., 2003). NC-specific deletion of Hand1 did not noticeably disrupt embryonic development, likely due to redundancy between Hand1 and Hand2 (Barbosa et al., 2007). Further removing one *Hand2* allele or deleting the pharyngeal arch expression of Hand2 in the *Hand1*$^{loxP/loxP}$; *Wnt1-Cre* (NC-null) background led to dysregulation of Pax9, Msx2, and Prrx2 in distal mesenchyme and hypoplasia of the pharyngeal arch-derived interdental mesenchyme and Meckel's cartilage, leading to a single fused lower incisor within a hypoplastic fused mandible (Barbosa et al., 2007).

In *Xenopus*, Hand1 marks cardiovascular precursors, but no expression is detected in NC derivatives at any stage, a striking difference compared to other vertebrates. In animal cap explants, Hand1 expression is strongly induced by ectopic expression of ventralizing signals Bmp2 and Bmp4 and by high doses of the cardiac muscle differentiation-inducing ActivinA (Sparrow et al., 1998). This activity is also observed in the chicken embryo, where Bmp4 treatment in the developing esophagus and gizzard induces Hand1 in non-NC-derived cells where it is not normally expressed (Wu and Howard, 2002).

3.3 Hand2

Hand2 is required for NC-derived structures including craniofacial cartilage and bone, the OFT of the heart, OFT cushions, and noradrenergic sympathetic ganglion neurons (Combs and Yutzey, 2009; Hendershot et al., 2008; Lee and Saint-Jeannet, 2009; Villanueva et al., 2002). Mice lacking Hand2 have a hypoplastic right ventricle and abnormal development of vessels arising from the heart and cell death of craniofacial pharyngeal arch precursors (Srivastava, 1999a).

3.3.1 Hand2 in the Pharyngeal Arches and Derivatives

Hand2, which is expressed in mesoderm- and NC-derived mesenchyme, has been identified as a key regulator of pathways necessary for pharyngeal arch morphogenesis (Clouthier et al., 2000; Ruest et al., 2003). NCCs entering the pharyngeal arches first migrate from the NT to the arches, proliferate, and finally undergo differentiation into terminal structures such as the mandibular skeleton and surrounding connective tissue (Clouthier et al., 2000). A *Hand2* pharyngeal arch enhancer driving expression of Cre recombinase activates a *ROSA26R*^{lacZ} reporter recapitulating pharyngeal arch expression. Labeled cells are first detected in postmigratory cranial NCCs within the pharyngeal arches and later in their derivatives (Ruest et al., 2003). In *Hand2*-null mouse embryos, the first and second arches are hypoplastic because of increased apoptosis and the third and fourth arches fail to form, but migration of NCCs appears normal. Msx1, one important downstream effector of Hand2, was undetectable in the mesenchyme of *Hand2*-null pharyngeal arches.

Interactions between the NC-derived mesenchyme and surrounding cells of the pharyngeal arches are crucial for development. An important early pathway controlling Hand2 expression in the distal first pharyngeal arch mesenchyme is the endothelin-1 (Edn1) signaling pathway. Mouse embryos mutant for Edn1, normally expressed in pharyngeal arch epithelia and the mesoderm-derived pharyngeal arch core (Fukuhara et al., 2004; Thomas et al., 1998), have a phenotype similar to *Hand2*-null embryos, and both Hand2 and Hand1 are downregulated in the pharyngeal and aortic arches of *Edn1*-null embryos. The *Hand2* locus contains an Edn1-dependent pharyngeal arch enhancer with four essential homeodomain binding sites. The homeodomain transcription factor Dlx6 binds these sites in an Edn1-dependent manner, and is downregulated in pharyngeal arches from EdnrA-null embryos (Charite et al., 2001; Fukuhara et al., 2004), demonstrating that Dlx6 acts as an intermediary between Edn1 signaling and Hand2 transcription (Charite et al., 2001). The G-protein-coupled receptor for Edn1, Ednra, is expressed in NC-derived mesenchyme. *Ednra*-null mice and *Dlx5*- or *Dlx6*-null mice are born with severe craniofacial defects resulting in neonatal lethality, including a homeotic transformation of mandibular-to-maxillary-like structures due to expansion of proximal first arch genes into the distal zones (Clouthier et al., 2000; Ruest et al., 2004). Constitutive activation of Ednra or misexpression of Hand2 in the Ednra domain induced the reverse: a transformation of maxillary-to-mandibular-like structures (Sato et al., 2008). Although NCC migration in *Ednra*-null embryos appears normal, there is complete loss or reduced expression of Hand2, Hand1, Goosecoid, Dlx2, Dlx3, Dlx6, and Barx1 in the postmigratory NC-derived arch mesenchyme, resulting in hypoplastic arches, differentiation defects, and apoptosis of some mesenchymal cells, much like the phenotype observed in *Hand2*-null embryos (Clouthier et al., 2000). A small distal Hand2 expression domain remains in *Ednra*-null mutants, and these cells likely contribute to formation of mostly normal lower incisors (Ruest et al., 2004). Mice deficient for the intracellular mediators of Edn1 signaling Gα(q) and Gα(11) also have reduced pharyngeal arch expression of Dlx3, Dlx6, Hand2, and Hand1, but not Msx1 (Ivey et al., 2003). The defects observed when any part of this Edn1-Ednra-Dlx6-Hand2 pathway is disrupted are reminiscent of those seen in human 22q11.2 deletion / velocardiofacial syndrome (Thomas et al., 1998).

By E9.5, Hand2 and Dlx6 expression in the mandibular mesenchyme is no longer dependent on the above-described endothelin signaling pathway and is at least partly mediated by FGF signals

(Fukuhara et al., 2004). In both mice and zebrafish, Hand2 expression in the pharyngeal arches also requires Mef2c (Miller et al., 2007; Verzi et al., 2007). Zebrafish *Mef2c* mutants have pharyngeal arch phenotypes resembling zebrafish *Edn1* partial loss-of-function mutants, and Mef2c is needed for expression of other genes downstream of Edn1 such as *Dlx5*, *Dlx6*, *Barx1*, and *Goosecoid*. Mef2c cranial NC expression does not require Edn1 signaling and Mef2c interacts genetically with Edn1, suggesting that Mef2ca functions within the Edn1 pathway (Miller et al., 2007). Hand2 and Runx2, a master regulator of osteogenesis, are partially overlapping in their expression domains in the developing mandible. Hand2, which is downregulated before osteoblast differentiation, negatively regulates intramembranous ossification of the mandible by physically interacting with Runx2 to inhibit Runx2 DNA binding and transcriptional activity. Mice with a hypomorphic mutation in *Hand2* have reduced mandibular mineralization and ectopic bone formation due to precocious osteoblast differentiation (Funato et al., 2009). A screen for genes dependent on Hand2 for expression identified *Ufd1l*, involved in degradation of ubiquitinylated proteins. Human *Ufd1l* maps to human 22q11 and was deleted in patients with 22q11 deletion syndrome, and mouse *Ufd11* is specifically expressed in those tissues affected in patients with 22q11 deletion (Yamagishi et al., 1999). The pathway including Ufd1l appears to be distinct from the Edn1 pathway, suggesting that Hand2 is involved in at least three parallel pathways needed for pharyngeal arch morphogenesis. An additional downstream target of Hand2 in the pharyngeal arches is Nebulette (Nebl), an actin-binding protein in the fetal heart, the human ortholog of which is deleted in DiGeorge syndrome 2 patients with cardiac and craniofacial abnormalities (Villanueva et al., 2002).

3.3.2 Hand2 in the Heart

In the developing heart, Hand2 is first expressed in mesoderm-derived cardiac precursors in the cardiac crescent and linear heart tube, becomes restricted to the developing right ventricle as heart looping occurs, and is expressed in the pharyngeal arch NC contributing to craniofacial structures and aortic arch arteries (McFadden et al., 2000). *Hand2*-null mouse embryos die between E9.5 and E10.5 (Hendershot et al., 2008; Srivastava et al., 1997), but some assessment of the role of Hand2 in early heart development can be made based on the mutant phenotype. Hand2 is required for the formation of both the mesoderm-derived right ventricle and the NC-derived aortic arch arteries; both tissues are hypoplastic in mutants (Srivastava et al., 1997; Srivastava, 1999b). A *Hand2-lacZ* transgene containing *lacZ* under the control of a 1.5-kb cardiac and NC-specific enhancer recapitulates endogenous Hand2 expression in the heart (McFadden et al., 2000). When Hand2 is deleted specifically in the NC (in *Hand2*$^{flox/flox}$; *Wnt1-Cre* embryos), embryos die at E12.5 with severe cardiovascular and facial defects and all NC-derived Hand2-expressing tissues seem to be affected (Hendershot et al., 2008; Morikawa et al., 2007). The early lethality in these embryos is due to loss of norepinephrine synthesis and can be rescued by activating adrenergic receptors (Morikawa and

Cserjesi, 2008). In rescued embryos, loss of Hand2 in the NC lineage leads to NC-related OFT and aortic arch artery defects including pulmonary stenosis, arch interruptions, retroesophageal right subclavian artery, and ventricular septal defects (Morikawa and Cserjesi, 2008). Hand2 functions in part by regulating signaling from the cardiac NC to other cardiac lineages but does not affect cardiac NC migration or survival. Loss of Hand2 in NC also indicates the role it plays within the cardiac NC to regulate differentiation and proliferation of the second heart field (SHF)-derived myocardium (Morikawa and Cserjesi, 2008). Hand2 also plays a role in preventing apoptosis of the SHF cells; an increase in ectopic SHF apoptosis in compound *Msx1*-null, *Msx2*-null embryos is associated with reduced expression of Hand1 and Hand2, although it is unclear whether Hand reduction occurs in the NC- or mesoderm-derived cells (Chen et al., 2007).

In humans, there may be a link between Hand2, which maps to human 4q33, and cardiovascular malformations commonly associated with terminal deletions of chromosome 4q, but strong evidence for this has not been demonstrated (Huang et al., 2002). In zebrafish, Hand2 expression is detected in the earliest precursors of all lateral mesoderm at early gastrula. Hand2 is later expressed in lateral precardiac mesoderm, pharyngeal arch NC derivatives, and posterior lateral mesoderm. At looping heart stages, cardiac Hand2 expression remained generalized with no apparent regionalization (Angelo et al., 2000). Knockdown of zebrafish dihydrofolate reductase (Dhfr), one of the enzymes needed for biological function of folic acid, results in cardiac defects similar to *Hand2* knockdown. Expression of Hand2 is reduced in Dhfr knockdown embryos, and cardiac defects are rescued by Hand2 overexpression (Sun et al., 2007). Disruption of *Hand* and *Mef2c* genes and a switch in MHC gene expression are early events in diabetic cardiomyopathy, likely due to the effects of oxidative stress on related signaling pathways (Aragno et al., 2006).

3.3.3 Hand2 in the Sympathetic Nervous System

Hand2 is expressed in several NC-derived tissues including components of the PNS. Hand2 is expressed in both the sympathetic and the parasympathetic divisions of the ANS (Dai et al., 2004). During human embryogenesis, Hand2 is expressed by sympathetic neuronal cells, extraadrenal chromaffin cells, and immature chromaffin cells in the adrenal gland, and is downregulated during chromaffin cell differentiation (Gestblom et al., 1999). In the chicken embryo, neuronal expression of Hand1 and Hand2 is restricted to sympathetic and enteric NC-derived ganglia (Howard et al., 1999; Howard et al., 2000). Sympathetic neurons are specified from NC precursors by a network of transcription factors including Hand2, Ascl1, Phox2a, and Phox2a. In the chicken embryo, overexpression of Hand2 also induces Gata2 expression (Tsarovina et al., 2004). Knockdown of Hand2 in NC-derived cells causes a significant reduction in neurogenesis and differentiation of catecholaminergic neurons but does not affect NT-derived neurons. *In vitro*, constitutive Hand2 expression is sufficient to induce catecholaminergic differentiation (Howard et al., 1999; Howard et al., 2000), and *in vivo,* Hand2 is sufficient to induce ectopic sympathetic neurons (Howard et al., 2000). Hand2 expression in chicken embryo sympathetic neurons is controlled by BMPs in vitro and in vivo and is induced downstream of Phox2b (Howard et al., 2000). In the mouse embryo, loss of Hand2 causes embryonic lethality by E9.5 and progressive loss of neurons and Th expression. Hand2 affects neural progenitor proliferation and expression of Phox2a and Gata3 (Hendershot et al., 2008). NC-specific deletion (*Hand2*$^{loxP/loxP}$*; Wnt1-Cre*) in the mouse results in lethality at E12.5 with severe cardiovascular and craniofacial defects, but NC-derived cells still populate regions of sympathetic nervous system (SNS) development and proliferate normally. Sympathetic precursors differentiate into neurons, demonstrating that Hand2 is not essential for SNS neuronal differentiation. However, the norepinephrine biosynthetic enzymes, Th, and dopamine β-hydroxylase (Dbh) were dramatically reduced in mutant embryos suggesting that the primary role of Hand2 in the SNS is determination of neuronal phenotype. Loss of Hand2 in the NC did not affect the expression of other genes regulating SNS development, including Phox2a, Phox2b, Gata2, Gata3, and Ascl1, but Hand2 was necessary for Hand1 expression (Morikawa et al., 2007). The major role of Hand2 during SNS development is most likely to permit sympathetic neurons to acquire a catecholaminergic phenotype (Morikawa et al., 2007). Hand2 acts together with Phox2a to regulate neurogenesis and noradrenergic marker gene expression. In these NC-derived cells, Hand2 and Phox2a are regulated by canonical and noncanonical BMP signaling. Protein kinase A activation by the noncanonical BMP pathway is needed to support neurogenesis and phosphorylation-dependent regulation of the *Dbh* promoter by Hand2. Activity of MapK is required for Hand2 transcription. Signaling affected by cAMP is necessary for regulation of *Dbh* promoter transactivation by Phox2a and Hand2 (Liu et al., 2005). MAP kinase activation and Bmp4-Smad1 signaling also mediate the differentiation of catecholaminergic neurons at least partly through induction of Hand2 (Liu et al., 2005; Wu and Howard, 2001). In zebrafish embryos with a Hand2 deletion (the *Hand2*$^{hands\ off}$ mutation), sympathetic precursor cells aggregate to form normal sympathetic ganglion primordia as marked by the expression of Phox2b, Phox2a, and Ascl1, but these cells have reduced expression of the noradrenergic marker genes Th and Dbh and the transcription factors Gata2 and AP-2α. By contrast, generic neuronal differentiation seems to be unaffected, showing an essential and specific function of Hand2 for noradrenergic differentiation of sympathetic neurons, and implicating AP-2α and Gata2 as downstream effectors (Lucas et al., 2006). Further illustrating the role of Hand2 in catecholaminergic differentiation, mesencephalic NCCs form cholinergic parasympathetic neurons in the ciliary ganglion, whereas trunk NCCs form both catecholaminergic and cholinergic neurons in sympathetic ganglia. Mesencephalic NCCs do not express the catecholaminergic transcription factor Hand2 in response to cranial BMP expression. Quail–chicken transplant experiments show that mesencephalic NCCs have a reduced capacity to undergo sympathetic differentiation; a subset of these cells never express Hand2 (Lee et al., 2005b).

3.3.4 Hand2 in the Enteric Nervous System

Hand2 is expressed in NC-derived precursors of enteric neurons and affects neurogenesis and specification of noradrenergic sympathetic ganglion neurons. Hand2 gain-of-function in NC progenitors results in increased neurogenesis and more neurons expressing vasoactive intestinal polypeptide (VIP). NC-specific deletion of Hand2 results in loss of all VIP-expressing gut neurons, reduction in Th-expressing neurons, abnormal Th-expressing neuron morphology, and abnormal patterning of the myenteric plexus (D'Autreaux et al., 2007; Hendershot et al., 2007). NC-derived cells are present, but neurons do not develop in explants from Hand2-null embryonic gastrointestinal (GI) tracts, and instead only glia are generated. Terminally differentiated enteric neurons do not develop after knockdown or conditional inactivation of Hand2 in migrating NC-derived cells (D'Autreaux et al., 2007). In the chicken embryo, Hand2 is expressed in neurons of the myenteric and submucosal ganglia. Hand2 is increased in NC-derived cells cocultured with the GI tract, suggesting a GI tract-derived factor regulates expression of Hand genes. Exposure of GI tract-derived NC-derived cells to Bmp4 significantly increases expression of Hand2 in all gut segments (Wu and Howard, 2002). Gdnf could not induce expression of Hand2 in NCCs but caused a modest increase in Hand2 in gut-derived NCCs from the esophagus and colon (Wu and Howard, 2002). In zebrafish, Hand2 is also required for normal development of the ENS and intestinal smooth muscle (Reichenbach et al., 2008).

3.3.5 Hand2 in Neuroblastoma

Neuroblastomas are tumors derived from the SNS that exhibit NC properties, perhaps as a result of impaired differentiation (Gestblom et al., 1999; Pietras et al., 2008). All examined neuroblastoma specimens and cell lines have detectable *Hand2* mRNA levels (Gestblom et al., 1999). In vitro studies from neuroblastoma samples show that Hand2 can homodimerize and forms heterodimers with another family of bHLH factors, the E proteins, and these heterodimers bind E-box DNA sequences (Dai and Cserjesi, 2002). High levels of another bHLH protein, Hif2α (a.k.a. Epas1), correspond to less-differentiated neuroblastoma cells and poorer prognosis. These high Hif2α-expressing cells tend to have higher expression levels of NC and early sympathetic progenitor marker genes such as *Hand2* (Pietras et al., 2008).

3.3.6 Hand2 Partners and Function

Transcriptional activity of Hand2 can be modulated depending on its phosphorylation state and its choice of heterodimer partner. Hand2 forms heterodimers with at least three E proteins: Tcf3, Tcf4, and Tcf12, and these bind to E-box elements with different affinities (Murakami et al., 2004b). Another partner of Hand2 is Jun activation domain-binding protein (JAB1/Cops5), which augments Hand2 transcriptional activity by enhancing binding through the HLH domain (Dai et al., 2004). Akt, a serine/threonine protein kinase involved in cell survival, growth, and differentiation, phosphorylates Hand2 and inhibits Hand2-mediated transcription (Murakami et al., 2004a).

3.4 Mitf

Microphthalmia transcription factor (Mitf) is a bHLH transcription factor critical for melanocytes and retinal pigment epithelia (RPE). Mitf is critical for melanocyte cell-fate choice during commitment from multipotent NC precursor cells. It is involved in differentiation, growth, and survival of pigment cells, interacting with several key signaling pathways, and is also likely involved in neoplastic growth of melanomas (Tachibana, 2000; Widlund and Fisher, 2003). In mammals, there are four described alternatively spliced isoforms of the single *Mitf* gene: the melanocyte lineage-enriched Mitf-M (the most abundant isoform), the RPE-enriched but mostly ubiquitous Mitf-A, the heart-enriched but mostly ubiquitous Mitf-H, and Mitf-C, expressed in many cell types, including RPE, but undetectable in melanocyte-lineage cells (Fuse et al., 1999; Tachibana, 2000). As described later in this section, fish and amphibians possess two distinct *Mitf* genes that encompass the roles of Mitf-M and Mitf-A. In the mouse, Mitf expression in the neuroepithelium and NC begins before the melanoblast marker Dct but is then coexpressed with Dct (Baker and Bronner-Fraser, 1997). Mitf-expressing cells coexpressing Dct and Kit migrate along the dorsolateral NC migration pathway (Opdecamp et al., 1997) and Mitf is gradually extinguished in all but hair follicle cells (Baker and Bronner-Fraser, 1997). Mitf is one of many white spotting genes associated with hypopigmentary disorders and deafness in several neurocristopathies (Hou and Pavan, 2008). More than 20 Mitf mutations have been identified in the mouse due to a range of NC-derived melanocyte deficiencies (Opdecamp et al., 1997; Steingrimsson et al., 1994). Loss of functional Mitf in mice results in complete absence of all pigment cells, resulting in microphthalmia and deafness (Yajima et al., 1999). $Mitf^{mi/mi}$ mouse loss-of-function mutants lack pigmentation, and are microphthalmic, whereas $Mitf^{vit/vit}$ mouse hypomorphic mutants display abnormal RPE pigmentation and progressive retinal degeneration (Gelineau-van Waes et al., 2008). The black-eyed white $Mitf^{mi\text{-}bw/mi\text{-}bw}$ mouse hypomorphic mutant has only reduced expression of Mitf-A and Mitf-H but complete loss of Mitf-M, resulting in a pigmented RPE but loss of melanocytes essential for body pigmentation and hearing (Yajima et al., 1999). In addition to melanogenesis, Mitf is required in vivo in a dosage-dependent manner for melanoblast survival during early migration away from the NT. $Mitf^{mi/+}$ embryos have fewer melanoblasts early, but later, the melanoblast population increases more rapidly to compensate, suggesting Mitf may also affect the rate of melanoblast increase during migration along the dorsolateral pathway (Hornyak et al., 2001). Expression levels of Mitf between melanocytes and RPE cells may differ, one possible mechanism for divergent fate specification. In cultured quail neuroretina cells exposed to excess Mitf, both melanocyte and RPE-like cells are induced. The expression level of Mitf is higher in melanocyte-like cells compared to RPE-like cells, suggesting that Mitf levels may in part determine the type of pigment cell induced. Overexpression of Mitf in cultured quail RPE cells causes these cells to develop into NC-like pigmented cells (Planque et al., 2004).

The development of melanophores in mammals and in fish shares a dependence on regulation by Mitf. Zebrafish have two *Mitf* genes, *Mitfa* and *Mitfb*. The two Mitf proteins encoded by these genes are homologous to distinct isoforms encoded by the single mammalian Mitf gene and play conserved yet divergent roles in zebrafish. Homozygous *Mitfa*^{nacre} mutants lack melanophores and early melanoblast markers throughout development but have increased numbers of iridophores (Lister, 2002; Lister et al., 2001; Lister et al., 1999). Mutations in the zebrafish Mitfa gene, expressed in all embryonic melanogenic cells, perturb only NC melanocytes, due to functional redundancy in RPE cells by Mitfb, which is not expressed in NC melanoblasts (Lister et al., 2001; Lister et al., 1999). In the NC, Mitfb can rescue melanophore development in *Mitfa*^{nacre} mutant embryos when transgenically expressed from the Mitfa promoter (Lister et al., 2001). Misexpression of Mitfa induces formation of ectopic pigmented cells in both wild-type and mutant embryos, demonstrating that Mitf is sufficient for pigment cell fate (Lister et al., 1999). In cultured ES-like cell lines from medaka, another fish species, the equivalent of Mitfa also seems to be sufficient for differentiation of pluripotent stem cells into melanocytes (Bejar et al., 2003). Besides fish, *Xenopus* also has two *Mitf* genes, *Mitfα* and *Mitfβ*, which are homologous to mouse *Mitf-M* (expressed specifically in the melanocyte lineage) and *Mitf-A* (strongly expressed in the RPE, and ubiquitous at lower levels). Mitfα is strongly expressed in the melanophore lineage (especially in premigratory melanoblasts) along with the developing RPE (Kumasaka et al., 2004).

3.4.1 Targets of Mitf

The main targets of Mitf during melanogenesis are Dct, Tyr, and Trp1, members of the tyrosinase gene family important for melanin synthesis. In embryos with severe Mitf mutations, NC-derived cells that would normally express Mitf lack Dct expression and soon disappear, showing that melanocyte development and survival require Mitf function. In contrast, neuroepithelial-derived Mitf-expressing cells of the retinal pigment layer are retained but are unpigmented. These cells express Dct, but not Tyr and Trp1 (Jiao et al., 2004; Nakayama et al., 1998). The murine *Dct* gene is expressed early in melanocyte development during embryogenesis, before Tyr and Trp1 (Jiao et al., 2004). The *Tyr*, *Trp1* and *Dct* promoters all contain a conserved E-box motif normally bound by Mitf. Mitf transactivates promoters of the tyrosinase gene family in both pigment cell lineages. These promoters also have specific DNA motifs corresponding to binding sites for RPE-specific factors such as Otx2 or melanocyte-specific factors Sox10 or Pax3 (Murisier and Beermann, 2006). A region of the *Dct* promoter critical for high Dct expression in melanocytes contains candidate binding sites for Sox10 and Mitf. Transfections into 293T and NIH3T3 cells show that Sox10 and Mitf can independently activate *Dct* expression, and when cotransfected, synergistically activate the *Dct* promoter (Jiao et al., 2004). Combinations of Pax3, Sox10, and Mitf regulate Dct transcription directly and synergistically (Jiao et al., 2006). Homozygous null *Mitf* mutants do not express Dct, and the number of cells expressing Kit is significantly reduced (Opdecamp et al., 1997). Wild-type NCC cultures rapidly give rise to cells expressing Mitf, Kit, and Dct, and with time, Kit expression increases concomitant with the appearance of pigmented cells. In contrast, cells cultured from Mitf mutant embryos did not increase Kit expression with time, suggesting Kit is downstream of Mitf. Mutant cells did not express Dct, never produced, and expression from Mitf itself was rapidly lost, suggesting a feedback mechanism. Mitf may initially play a role in the transition of precursor cells to melanoblasts and later regulating melanoblast survival by controlling Kit expression (Opdecamp et al., 1997). Mitf is required for maintenance of Kit expression in melanoblasts and Kit signaling in turn modulates Mitf activity and stability in melanocyte cell lines (Hou et al., 2000). In primary NCC cultures, initiation of Mitf expression in melanoblasts does not require Kit. A small number of Kit-null Mitf-positive cells can be maintained for at least 2 weeks in culture; these cells express several pigment cell-specific genes activated by Mitf (Dct, Silver(Si), and Trp1), but lack expression of Tyr, which encodes the rate-limiting enzyme in melanin synthesis. This demonstrates that the presence of Mitf alone is not sufficient for Tyr expression in melanoblasts. However, elevation of cAMP levels in these cultures increases the number of Mitf-positive cells, induces Tyr expression, and results in differentiation of melanoblasts into mature, pigmented melanocytes (Hou et al., 2000).

Mitf mutant embryos also have reduced expression of the membrane-associated transporter protein (Matp/Slc45a2) (Baxter and Pavan, 2002). Matp expression was similar to the melanogenic enzymes Dct and Trp1, suggesting similar regulation, and *Matp* mutations result in pigment alterations in mice (*underwhite/uw* mutants), fish (medaka b-locus), and humans (Oculocutaneous Albinism Type 4). Further evidence suggesting Matp acts in a pigment cell autonomous manner is the expression of mouse Matp in presumptive RPE starting at E9.5, and in NC-derived melanoblasts starting at E10.5 (Baxter and Pavan, 2002). Additionally, Mitf mutant embryos lose expression of the melanosomal gene Si, also normally expressed in the developing RPE starting at E9.5, and in NC-derived melanoblasts starting at E10.5. Si expression in dorsal regions precedes Dct expression, suggesting Si is an earlier melanoblast marker (Baxter and Pavan, 2003).

3.4.2 Regulation of Mitf by Sox10 and Pax3

Sox10 can transactivate the *Mitf* promoter up to 100-fold by binding to a region conserved between mouse and human *Mitf* promoters. A dominant-negative Sox10 mutant reduces wild-type Sox10 induction of Mitf (Potterf et al., 2000). Both Sox10 and Pax3 directly regulate *Mitf*, and the strong transactivation of *Mitf* by Sox10 is further stimulated by synergy with Pax3 as each protein binds independently (Lang and Epstein, 2003; Potterf et al., 2000). At E11.5, mouse embryos homozygous for the *Sox10*^{Dom} mutation entirely lack NC-derived cells expressing the lineage markers Kit, Mitf, or Dct. In *Sox10*^{Dom} heterozygous embryos, melanoblasts expressing Kit and Mitf do occur, albeit in reduced numbers, and pigmented cells eventually develop in almost normal numbers in culture and in vivo. Dct is not expressed, suggesting that Sox10 acts as a critical transactivator of Dct (Potterf et al., 2001). Melanophore defects in zebrafish *Sox10*^{colourless} mutants can be explained by disruption of Mitf expression. *Sox10*^{colourless} NCCs adopt mesenchymal fates, and NCCs that would normally adopt nonmesenchymal fates generally fail to migrate, do not differentiate, and instead undergo apoptosis. All defects of affected NC derivatives are consistent with a primary role for Sox10 in specification of nonmesenchymal NC derivatives (Dutton et al., 2001). Mitfa expression is undetectable in Sox10-null zebrafish. The zebrafish *Mitfa* promoter contains Sox10 binding sites necessary for activity in vivo and in vitro, consistent with studies in mammalian cell cultures demonstrating Sox10 directly regulates *Mitf*. Reintroduction of Mitfa expression in NCCs can rescue melanophore development in *Sox10*-null embryos in a manner quantitatively indistinguishable from rescue in *Mitfa*-null embryos, suggesting the essential function of Sox10 in melanophore development is transcriptional regulation of *Mitfa* (Elworthy et al., 2003). A melanocyte-specific *Mitf* enhancer contains binding sites for Sox10. Sox10 and Mitf-M are coexpressed in melanoblasts migrating toward the otic vesicle of mouse embryos but are separately expressed in different cell types of the newborn cochlea suggesting that Sox10 regulates transcription from the *Mitf* promoter in a developmental stage-specific manner (Watanabe et al., 2002a). *Mitf* regulation by Sox10 is disrupted by sumoylation of Sox10, which represses Sox10 transcriptional activity on *Mitf* (Girard and Goossens, 2006). In the melanocyte lineage, Pax3 regulates Mitf and Trp1 (Corry and Underhill, 2005).

Mitf is also activated by α-melanocyte-stimulating hormone (Msh/Pomc) through a conserved cAMP response element (CRE). cAMP-mediated CRE-binding protein activation of the *Mitf* promoter requires a second nearby DNA element that is bound and activated by Sox10. In NC-derived melanoma and neuroblastoma cells lacking Mitf-M and Sox10 expression, *Mitf* promoter responsiveness to cAMP depends on Sox10, and Sox10 transactivation is likewise dependent on the CRE. Ectopic Sox10 expression, in cooperation with cAMP signaling, activates the *Mitf* promoter and expression of endogenous *Mitf* transcripts in neuroblastoma cells. This activation does not occur with the *Sox10*^{Dom} allele (Huber et al., 2003). High-level activation of cAMP signaling attenuates Bmp2-induced sympathoadrenal cell development and induces melanogenesis by inducing transcription of *Mitf*. Dominant-negative Creb1 inhibits Mitf expression and melanogenesis, suggesting Creb1 activation is necessary for melanogenesis. However, constitutive activation of Creb1 without PKA activation is insufficient for Mitf expression and melanogenesis, indicating

PKA regulates additional aspects of *Mitf* transcription (Ji and Andrisani, 2005). In vitro studies demonstrate that the phorbol ester 12-tetradecanoylphorbol 13-acetate (TPA) is another factor that induces NCC differentiation into melanocytes and stimulates proliferation and differentiation of melanocytes. In primary NCC cultures, TPA is necessary for Mitf upregulation and melanin synthesis. In an immortalized melanocyte cell line that proliferates in the absence of TPA, TPA significantly increases the mRNA levels of the tyrosinase gene family (*Tyr*, *Tyrp1*, and *Dct*) and the expression of Mitf (Prince et al., 2003). When overexpressed in melanocytes, Emx1 and Emx2 downregulate *Mitf*, *Tyrp1*, *Dct*, and *Tyr*, suggesting that these genes may restrict expression of Mitf by inhibiting activation in neuroepithelial derivatives other than melanocytes (Bordogna et al., 2005).

3.4.3 Regulation of Zebrafish Mitf by Foxd3

The zebrafish *Hdac1*^{colgate} mutant has increased and prolonged expression of Foxd3, resulting in reduced numbers, delayed differentiation, and decreased migration of NC-derived melanophores and their precursors due to a severe reduction in the number of Mitfa-positive melanoblasts. Hdac1 is required to suppress Foxd3 expression in the NC, thus de-repressing Mitfa resulting in melanogenesis in a subset of NC-derived cells (Ignatius et al., 2008). Foxd3 acts directly on the *Mitfa* promoter to negatively regulate Mitfa expression (Curran et al., 2009). Mitfa is only expressed in a Foxd3-negative subset of NCCs, and Foxd3 mutants have more cells expressing Mitfa. Foxd3 prevents a subset of NC-derived precursors from acquiring a melanophore fate by repressing the *Mitfa* promoter and is not expressed in terminally differentiated melanophores. Foxd3 is, however, expressed in xanthophore precursors and iridophores (Curran et al., 2009).

3.4.4 Mitf Downstream of Wnt Signaling

In cultured human melanocytes, exogenous Wnt3a upregulates expression of Mitf. Wnt3a signaling likely recruits β-catenin and Lef1 to the Lef1-binding site of the Mitf promoter (Takeda et al., 2000). In zebrafish, Wnt signals are necessary and sufficient for NCCs to adopt pigment cell fates, which also require Mitf. As in mammalian cells, a promoter region of zebrafish *Mitfa* containing Tcf/Lef binding sites mediates Wnt responsiveness. An *Mitf* reporter construct is strongly repressed by a dominant-negative Tcf in melanoma cells. Mutation of Tcf/Lef sites abolishes Lef1 binding and reporter function in vivo (Dorsky et al., 2000). Functional cooperation of Mitf with Lef1 results in synergistic transactivation of Dct, an early melanoblast marker. β-Catenin is required for efficient transactivation but is dispensable for the interaction between Mitf and Lef1. The interaction with Mitf is unique to Lef1 and not detectable with Tcf1. Lef1 also cooperates with Mitf-related proteins, such as the bHLH factor Tcfe3, to transactivate the *Dct* promoter (Yasumoto et al., 2002).

Mitf functions downstream of the canonical Wnt pathway in mammalian melanocyte and zebrafish melanophore development and is regulated by β-catenin across species (Widlund et al., 2002). β-Catenin is expressed throughout development of the melanocyte lineage and contributes to regulation of Mitf expression (Larue et al., 2003). The Wnt/β-catenin canonical signaling pathway also plays a role in melanoma. Several components of the Wnt pathway are altered in melanoma tumors and cell lines. Activated β-catenin is found in approximately 30% of human melanoma nuclei and can induce genes regulating proliferation (*Myc* or *CyclinD1*) in addition to regulating cell lineage-restricted genes (*Brn2*) and melanocyte-specific genes (*Mitf* and *Dct*) (Larue and Delmas, 2006).

3.4.5 Posttranslational Regulation of Mitf

Mitf is posttranslationally regulated by phosphorylation and ubiquitination affecting protein activity and stability (Galy et al., 2002). Kit signaling is linked to phosphorylation of Mitf through activation of MAP kinase and Rps6ka1 (ribosomal protein S6 kinase), and IGF-1 and HGF/SF signaling may also be involved. Phosphorylation of Mitf is carried out by GSK3β (Tachibana, 2000). Mitf is also modified in melanoma cells by sumoylation: mutations affecting sumoylated residues, although displaying normal DNA binding, stability, and nuclear localization, result in a substantial increase in the transcriptional stimulation of promoters containing multiple (but not single) Mitf binding sites and enhanced cooperation with Sox10 on the *Dct* promoter. Sumoylation of Mitf regulates the protein's transcriptional activity with respect to synergistic activation and sumoylation plays a significant role among the multiple mechanisms regulating Mitf (Murakami and Arnheiter, 2005). In melanocyte cultures, the MAPK pathway targets Mitf for ubiquitin-dependent proteolysis, resulting in a rapid degradation and downregulation (Galy et al., 2002).

3.4.6 Mitf in Human Disease

Mutations in human *Mitf* are found in patients with type 2 Waardenburg syndrome (WS2), a dominantly inherited syndrome associated with hearing loss and pigmentary disturbances, and sometimes occurring in conjunction with other NC syndromes (Boissy and Nordlund, 1997; Lalwani et al., 1998; Steingrimsson et al., 1994; Tachibana, 2000; Van Camp et al., 1995). WS2 is heterogeneous; 15–20% of cases are caused by mutations in Mitf, particularly affecting Mitf-M (Morell et al., 1997; Read and Newton, 1997; Tassabehji et al., 1995; Yajima et al., 1999). The dominant melanophore phenotype in type 4 Waardenburg syndrome individuals with Sox10 mutations likely results from failure to activate Mitf in a normal number of melanoblasts (Elworthy et al., 2003). In human melanoma, many pigmentation markers are lost, but Mitf expression is still active, even in unpigmented tumors, suggesting a role for Mitf beyond differentiation. Many primary human melanomas have aberrant nuclear accumulation of β-catenin, a potent growth mediator of melanoma cells dependent on Mitf, which is a downstream target of β-catenin. The suppression of melanoma clonogenic growth that occurs when β-catenin-Tcf/Lef regulation is disrupted can be rescued by constitutive expression of Mitf, suggesting a prosurvival mechanism for Mitf. β-Catenin regulation of Mitf represents a tissue-restricted pathway that significantly influences the growth and survival of melanoma cells (Widlund et al., 2002). Key proteins in melanogenesis such as Mitf-M, Sox10, Pax3, Trp1, and Tyr are absent or greatly reduced in the bulbs of graying hair compared to black hair. Hair graying is likely caused by defective migration of melanocyte stem cells into the bulb area of hair (Choi et al., 2008).

3.5 Myc

Myc (also known as c-Myc, a protooncogene), is a bHLH transcription factor generally important for control of cell cycle progression and proliferation. In the NC, Myc plays a role in specification of NCCs and maintenance of a proliferative, undifferentiated NC progenitor state (Barembaum and Bronner-Fraser, 2005; Hong et al., 2008d; Light et al., 2005). In *Xenopus*, Myc is expressed at the neural plate border before many early NC markers and serves as an essential early regulator of NCC formation. Knockdown of Myc results in the absence of NC precursor cells and later derivatives (Barembaum and Bronner-Fraser, 2005; Bellmeyer et al., 2003; Hong et al., 2008d). Forced expression of Slug or Twist, which require Myc for normal expression, compensates for defects caused by loss of Myc (Rodrigues et al., 2008). In the mouse and zebrafish, Myc is involved in craniofacial development, and at least in zebrafish, craniofacial defects resulting from Myc knockdown are a result of increased NCC death (Hong et al., 2008d; Wei et al., 2007). There are also several lines of evidence supporting a role for Myc in maintenance of NCSC properties. One target of Myc is Id3, a factor also required for formation and maintenance of NCSCs (Light et al., 2005). Expression of an inducible Myc-ER^{T} fusion protein driven from the ubiquitously expressed *ROSA26* locus (the ROSAMER mouse strain or the JoMa1.3 cell line) in primary NC cultures maintains NCSC proliferative capacity and differentiation potential (Jager et al., 2004; Maurer et al., 2007). NCSCs such as those derived from the adult palatum express Myc and a complement of other stem cell markers and reprogramming factors (Widera et al., 2009).

3.6 Mycn

Mycn is expressed mainly in the nervous system, including components of the NC-derived PNS, such as the dorsal root ganglia (DRGs) (Wada et al., 1997). In the PNS, Mycn controls proliferation and differentiation of progenitor cells (Grimmer and Weiss, 2006; Kobayashi et al., 2006) and is quickly extinguished as these progenitors undergo terminal differentiation and stop proliferating (Thomas et al., 2004). In the chicken embryo, Mycn is expressed in all early NCCs and is then extinguished during migration to the ganglion and nerve cord regions in all cells except those undergoing neuronal differentiation (Wakamatsu et al., 1997). Overexpression of Mycn in chicken NC culture promotes neuronal differentiation and ventral migration of NCCs (Wakamatsu et al., 1997). In the mouse, overexpression of Mycn does not alter the number of multipotent NCCs or affect differentiation toward the glial lineage, but, as a result of cell cycle reentry, increases proliferation and apoptosis in neuronal cells of the DRG, resulting in fewer total cells in the DRGs (Kobayashi et al., 2006). Mycn overexpression can also alter neuronal subtype fate choice; proportions of proprioceptive neurons are increased significantly in mice overexpressing Mycn (Kobayashi et al., 2006).

In addition to the role of Mycn in maintaining the undifferentiated state of migrating NC neuroblasts, Mycn is associated with metastatic disease (Wada et al., 1997). Overexpression of Mycn can transform cells in vivo and in vitro (Thomas et al., 2004). Between 20% and 38% of neuroblastomas, tumors that share characteristics of embryonic NCCs, have *Mycn* amplification (Ngan et al., 2007) correlating with aggressive tumors and poor prognosis (Grimmer and Weiss, 2006; Koppen et al., 2007; Ohira et al., 2003; Terui et al., 2005). Neuroblastic-like and stem cell-like sublines have high Mycn expression, whereas nonneuronal NC-like precursors have much lower Mycn and significantly decreased malignant potential, directly proportional to Mycn expression (Spengler et al., 1997). In neuroblastomas, Mycn may affect function of several important signaling pathways. In neuroblastoma cell lines, Mycn represses the Wnt signaling inhibitor Dickkopf-1 (Dkk), known to function in NCC migration (Koppen et al., 2007). Mycn levels are also influenced by Shh and PI3K signaling (Grimmer and Weiss, 2006). Extra copies of *Mycn*, occurring frequently within neuroblastoma cell lines, will transactivate Id2, resulting in Id2 overexpression, which in turn inactivates the tumor suppressor Rb, allowing cell cycle progression (Lasorella et al., 2000). Some human neuroblastoma cell lines can be induced to undergo differentiation into various NC-derived cell types upon treatment with retinoic acid (RA), and downregulation of Mycn by RA and other molecules such as interferon-γ is a key step in differentiation of neuroblastoma cells (Cinatl et al., 1996; Wada et al., 1997). Forced Mycn overexpression renders neuroblastoma cells insensitive to RA (Giannini et al., 1999).

3.7 Tcf3, Tcf4, AND Tcf12 (E PROTEINS)

The E proteins are a family of class A bHLH transcription factors that bind E-box promoter elements typically as heterodimers with class B bHLH transcription factors. The E proteins Tcf3 (previously known as E2A, E12, or E47), Tcf4 (E2.2), and Tcf12 (Alf1, Me1) interact with the NC-enriched bHLH factor Hand2 as heterodimers that bind E-box elements. Each heterodimer binds with different affinity, giving each combination a unique specificity. E-protein genes have broad expression patterns, so it is presumably the ability to form unique heterdimers that confers tissue-specific activities (Dai and Cserjesi, 2002; Murakami et al., 2004b). Tcf3 also forms heterodimers with Hand1 (Cserjesi et al., 1995). Tcf3 plays a role in repression of *E-cadherin* expression and EMT (Perez-Moreno et al., 2001), and it also induces expression of the cell cycle regulator p21, but this expression can be inhibited by Snail or Snail and Twist together (Takahashi et al., 2004). In hypoxic neuroblastoma cells, Tcf4, a dimerization partner for the proneuronal lineage-specifying bHLH transcription factors Neurod1 and Neurog1, was downregulated, whereas inhibitors of neural differentiation Id2 and Hes1 were induced, suggesting a mechanism for hypoxia-induced dedifferentiation of neuroblastoma cells (Jogi et al., 2004). Tcf3 and Tcf4 can also form heterodimers with the inhibitory Id proteins, preventing functional hetero- or homodimeric DNA binding complexes (Jogi et al., 2002).

3.8 Twist1

Members of the *Twist* gene family encode for basic helix–loop–helix transcription factors. In mammals, there are two Twist family members, Twist1 and Twist2 (also known as Dermo1), and only Twist1 expression has been reported in NCCs. In zebrafish, in addition to a duplicated *Twist1* (twist1a and twist1b) and a single *Twist2*, there is a related *Twist3* gene. Twist1 is a master regulator of EMT in the embryo and during cancer metastasis. Twist1 function can be regulated by partner choice, and phosphorylation of residues within the bHLH domain alters partner affinities for Twist1 (Firulli and Conway, 2004; Firulli and Conway, 2008).

3.8.1 Twist1 in the Cranial Mesenchyme and Pharyngeal Arches

In the mouse, Twist1 is first expressed in extraembryonic tissues, then within some ectodermal cells of the primitive streak, and subsequently in mesoderm outside the primitive streak (Fuchtbauer, 1995; Stoetzel et al., 1995). Beginning around E8–E8.25, Twist transcripts accumulate in the cranial mesenchyme in both cranial NCCs and NC in the pharyngeal arches (Fuchtbauer, 1995; Gitelman, 1997; Stoetzel et al., 1995). During embryogenesis, Twist1 persists in mesodermal and NCC derivatives, often at sites of epithelial–mesenchymal interactions, and most prominently in progenitor cells of the muscle and cartilage/skeletal lineages, suggesting a role for Twist in inhibiting differentiation in these cell types (Fuchtbauer, 1995; Stoetzel et al., 1995). *Twist1*-null mouse embryos die at E11.5 with defects in the cranial mesenchyme; the cranial neural folds do not fuse as a result of cell-autonomous Twist1 function in cranial mesenchyme (Chen and Behringer, 1995; Gitelman, 1997). Malformation of the pharyngeal arches and defects in the somites and limb buds are also observed (Chen and Behringer, 1995; Soo et al., 2002). In the cranial mesenchyme, cells lacking Twist1 migrate abnormally and sometimes lack mesenchymal characteristics, suggesting Twist1 regulates migration and specification of the cranial mesenchymal cells needed for fusion of the cranial neural folds (Chen and Behringer, 1995; Gitelman, 1997). In the first pharyngeal arch,

Twist1 is required in both the mesoderm-derived cranial mesenchyme for directing early migration of NCCs and in the NCCs themselves; without Twist1, NCCs stray from their normal migratory path (Soo et al., 2002; Vincentz et al., 2008). Twist1 is also necessary for differentiation of the first pharyngeal arch tissues into bone, muscle, and teeth (O'Rourke and Tam, 2002; Ota et al., 2004; Soo et al., 2002). Consistent with this, mutations in human Twist contribute to syndromic craniofacial abnormalities and perhaps also to nonsyndromic cleft lip and palate, one of the most common human birth defects (Bacon et al., 2007).

In addition to its requirement for NC emigration from the NT and migration into the pharyngeal arches, Twist1 is also required for later migration along cardiac NCC pathways. In *Twist1*-null mice, cardiac NCCs are delayed in their colonization of the endocardial OFT cushions, which later contain abnormal, nodular mesenchyme derived exclusively from the cardiac NC. A subpopulation of *Twist1*-null NCCs that successfully migrate to the OFT has defects in maturation, migration, and adhesion. There are similar nodules in pharyngeal arches, and the dorsal NT has an expanded domain of NCCs (*Wnt1-Cre*-lineage marked cells) (Vincentz et al., 2008). Loss of Twist1 also affects patterning of the cranial ganglia but not patterning of the peripheral ganglia derived from trunk NC (Ota et al., 2004).

3.8.2 Twist1 in the Skull Vault

The frontal and parietal bones that form the vertebrate skull vault are composed of cells of NC and mesodermal origin, respectively. These two divergent cell lineages come together at positions between the ectoderm and cerebral hemispheres. The boundary between these two mesenchymal cell populations becomes the developing coronal suture, a major growth center for the skull, where both lineages proliferate and differentiate toward an osteogenic fate (Ishii et al., 2003; Merrill et al., 2006). Craniosynostosis (also termed *coronal synostosis*, and for a less common subset, *Saethre-Chotzen syndrome*) is a relatively common human birth defect resulting from premature fusion of the developing frontal and parietal bones (Merrill et al., 2006). One cause of human craniosynostosis, calvarial foramina (persistent unossified areas within the skull vault), is linked to mutations in *Twist1*, *Msx2*, and the gene encoding the ephrin A4 ligand, *Efna4* (Firulli and Conway, 2008; Ishii et al., 2003; Merrill et al., 2006; Paznekas et al., 1999), three factors that cooperate in regulation of skeletogenic mesenchyme proliferation and differentiation in the developing frontal bone. Heterozygous loss of murine Twist1 results in a defective boundary between frontal NC and parietal mesoderm mesenchymal cells, allowing the NCCs to invade the undifferentiated mesoderm of the coronal suture (Ishii et al., 2003; Merrill et al., 2006). This is accompanied by an expansion in Msx2 expression and reduction of Efna2, Efna4, and Epha4; Efna4 expression in *Twist1*-heterozygous embryos can be rescued by loss of one allele of Msx2, which in turn rescues the craniosynostosis defect (Merrill et al., 2006). In *Twist1*-null, *Msx2*-null double mutants, the quantity and proliferation of frontal bone skeletogenic mesenchyme are reduced compared to individual mutants. Msx2 and

Twist1 likely cooperate in regulation of skeletogenic mesenchyme differentiation and proliferation, with each independently controlling NCC and mesoderm contribution (Ishii et al., 2003). Mice deficient for Efna4 exhibit defects in the coronal suture NC–mesoderm boundary much like *Twist1* heterozygous mice, and compound *Twist1* heterozygous, *Epha4* heterozygous mice have more severe defects than those of individual heterozygotes, indicating a genetic interaction. DiI labeling of migratory osteogenic progenitors contributing to frontal and parietal bones reveals that Twist1 and Epha4 are required for exclusion of these cells from the coronal suture (Ting et al., 2009).

3.8.3 Twist in Zebrafish and *Xenopus*

Zebrafish *Twist* gene expression is detected in cranial NC, sclerotome, lateral plate mesoderm, and in the dorsal aorta (Germanguz et al., 2007). During early development, Twist1 is expressed in cranial mesenchyme and later in NC of the pharyngeal arches, consistent with a role in craniofacial development (Yeo et al., 2007). In *Xenopus*, Twist1 is expressed at the time of early NC specification together with Snail and Slug in the presumptive neural folds and marks the presumptive NC (Hopwood et al., 1989; Linker et al., 2000). Twist1 expression follows expression of Snail and Slug and can be induced by Snail during NC specification (Aybar et al., 2003; Linker et al., 2000).

3.8.4 Twist as an NC Marker and in Cancer and NC-Derived Stem Cells

Knockdown of Rhov, a Ras homolog expressed in early NC and essential for NC induction, disrupts expression of Twist1, Slug, and Sox9 (Guemar et al., 2007). Twist1 is also expressed in human dental NC-derived progenitor cells (Degistirici et al., 2008), porcine NC skin-derived precursors (pSKPs) (Zhao et al., 2009), and multipotent rat periodontal ligament-derived NC stem cells (Techawattanawisal et al., 2007). Twist1 and Sox9 are overexpressed in NC Schwann cell-derived malignant peripheral nerve sheath tumors (MPNSTs). Reduction of Twist1 (implicated in metastasis, chemotherapy resistance, and inhibition of apoptosis) in MPNST cells has no effect on apoptosis or chemotherapy resistance but results in inhibition of MPNST cell chemotaxis (Miller et al., 2006). During carcinoma of the parathyroid glands, which have a dense outer mesenchymal NC-derived component, expression of Twist1 (and Snail) changes from a homogenous distribution among cancer cells to localization in the cells at the invasive front of the cancer with a corresponding loss of membranous Cdh1 (E-cadherin) (Fendrich et al., 2009).

3.9 OTHER BASIC HELIX–LOOP–HELIX GENES: Usf1, Usf2, Tcfe3, Mnt, Olig3

The upstream stimulatory factors Usf1 and Usf2 are bHLH/leucine zipper transcription factors. During *Xenopus* embryogenesis, Usf1 and Usf2 are highly expressed in NC and neural tissues, eye, and otic vesicle (Fujimi and Aruga, 2008). In primary cultures of rat aortic vascular smooth muscle cells (VSMCs), Usf1 and Usf2 homo- and heterodimers bind an essential E-box element in the promoter of the aortic preferentially expressed gene-1 (Apeg1/Speg), expressed highly in differentiated VSMCs, some of which are NC-derived. Dominant-negative Usf mutant proteins repress *Apeg1* promoter activity, and Usf1 alone transactivates the Apeg1 promoter (Chen et al., 2001). Tcfe3, an Mitf-related bHLH protein, cooperates with Lef1 to transactivate the Dct promoter in NC derivatives (Yasumoto et al., 2002). The bHLH Max-binding protein Mnt is present in *Xenopus* migrating NCCs, but its function is unexplored (Juergens et al., 2005). In mouse, chicken, and zebrafish, Olig3 plays an essential role in establishing the boundary between NC and the lateral neural plate (Filippi et al., 2005). Knockdown of zebrafish Olig3 increases NCCs at the expense of interneurons and astrocytes from the lateral neural plate. Disruption of BMP or Notch signaling and loss of NCCs cause upregulation/expansion of the Olig3 expression domain; Olig3 repression rescues the NC loss (Filippi et al., 2005).

• • • •

CHAPTER 4
ETS Genes

The *Ets* gene superfamily encodes a class of transcription factors that bind a purine-rich DNA sequence through an 85-amino-acid ETS domain (Dhordain et al., 1995). This family consists of many members with roles in the NC: Ets1, Ets2, Etv1, Etv4/Pea3, Etv5/Erm, Erg, Fli1, and Elk1. The *Ets1* and *Ets2* genes were identified by their sequence homology to the v-ets oncogene of the E26 avian erythroblastosis virus (Maroulakou et al., 1994). Many Ets genes are expressed in migratory cells such as NC, endothelial, and pronephric duct cells, and other embryonic areas affected by EMTs. *Ets* genes influence cell specification (Remy and Baltzinger, 2000) and appear to coordinate changes in adhesion molecule expression and degradation of the ECM, suggesting an important role in regulating cell motility (Meyer et al., 1997; Remy and Baltzinger, 2000; Tahtakran and Selleck, 2003). Because of these properties, these genes are often involved in tissue remodeling during embryogenesis, wound healing, and metastasis (Remy and Baltzinger, 2000).

4.1 Ets1 AND Ets2

In the mouse, Ets1 expression is detected during a narrow developmental stage in the developing nervous system, including the presumptive hindbrain regions, NT, NC, and in the first and second pharyngeal arches. Ets2 expression is limited to the developing limb buds and distal tail. Later, *Ets1* mRNA is observed in developing vascular structures, including the heart, arteries, capillaries, and meninges, whereas *Ets2* mRNA is restricted to developing bone, tooth buds, epithelial layers of the gut, nasal sinus and uterus, and several regions of the developing brain (Maroulakou et al., 1994). In the chicken embryo, Ets1 expression is transiently induced in epithelial structures, including during emigration of NCCs (Fafeur et al., 1997). Chicken Ets1 is upregulated in the cranial neural folds and dorsal neural tube approximately 4–6 h before commencement of NC migration and continues to be expressed in migrating cranial NCCs and later in some NC-derived tissues (Tahtakran and Selleck, 2003). Chicken Ets1 influences cranial NCC mobilization, and although it does not promote EMT on its own, it can augment the effect of Slug on EMT initiation (Theveneau et al., 2007). Mafb, expressed in r5 and r6 and NC derived from these rhombomeres, interacts with chicken Ets1 (Eichmann et al., 1997). In *Xenopus* neurula and tailbud stages, *Ets1* and *Ets2* transcripts are detected in NCCs and their derivatives. *Xenopus* Ets1 is an NC marker downstream of

Snail that is induced at the time of NC specification, and activation by Snail occurs in whole embryos and in animal caps (Aybar et al., 2003). Consistent with a role in EMT, human Ets1 expression is associated with invasive processes in tumors (Fafeur et al., 1997). Ets1 plays a role in tumor vascularization and invasion by regulating expression of matrix-degrading proteases in endothelial cells and fibroblasts in the tumor stroma. Ets1 is expressed in migrating NCCs from which melanocytes arise and is upregulated both in vivo and in vitro in malignant melanoma compared to benign melanocytic lesions and primary melanocytes. Knockdown of Ets1 in the melanoma cell line Mel Im correlates with reduced expression of several Ets1 target genes involved in invasion, such as *Mmp1*, *Mmp3*, *Plau*, and *Itgb3*, and reduced invasive capacity (Rothhammer et al., 2004). Ets1 is associated with tumor progression in various carcinomas and malignant melanoma and is variably expressed in most primary melanomas, in normal benign melanocytes, and all nevi. Ets1 expression is lower in primary melanomas than in common nevi, and metastatic melanomas expressed significantly less Ets1 than primary melanomas (Torlakovic et al., 2004). Snail, upregulated in human melanoma cells, has a conserved 3′ enhancer with overlapping Ets and Yy1 consensus sequence required, together with a Sox10 binding sequence, for full enhancer activity (Palmer et al., 2009).

4.2 Elk1 AND Elk4

Elk1, another member of the Ets oncogene family, is downregulated together with Klf4 and Myocardin by miR-145 and miR-143 to promote differentiation and repress proliferation of VSMCs, including NC-derived VSMCs (Cordes et al., 2009). *Xenopus* Elk4 (Sap1) encodes a member of

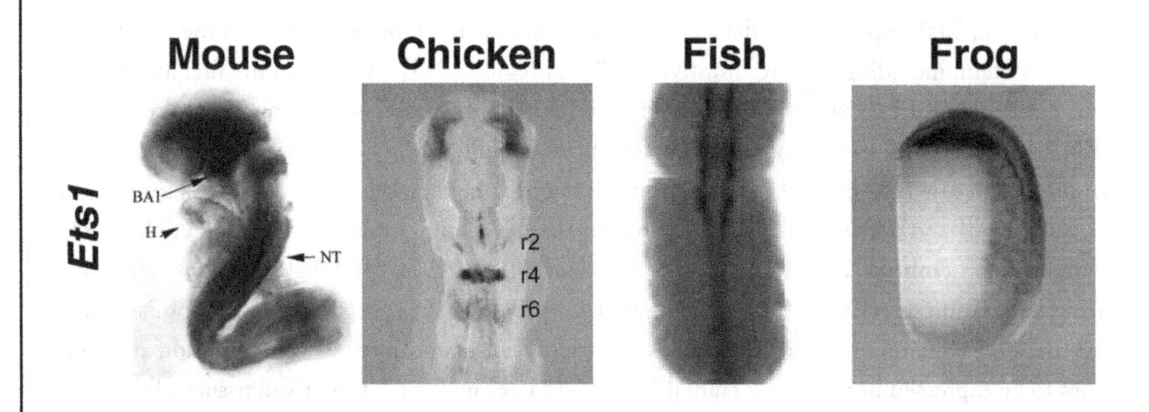

FIGURE 4.1: Whole-mount in situ hybridizations show Ets1 in mouse, chicken, fish, and frog embryos. (Mouse: Ye et al., 2010, 8.75 dpc embryo, dorsal view, anterior at top; Chicken: Theveneau et al., 2007, 13-somite stage embryo dorsal view; Zebrafish: Pham et al., 2007, 15 somite embryo, reconstructed/extended dorsal view; Xenopus: Sasai et al., 2001, stage 23 embryo, lateral view, anterior at top).

the ternary complex factor subfamily of ETS transcription factors. During neurulation, Elk4 is expressed in cells participating in neural tube formation, in the sensorial layer of the epidermal ectoderm, and in an anterior region of the ventral mesoderm. Later, Elk4 expression is observed in the eye, otic vesicle, pharyngeal arches, heart, pronephros, somites, and the developing CNS (Nentwich et al., 2001).

4.3 Etv1, Etv4, AND Etv5

Etv subfamily ETS-domain proteins play important roles in regulating transcriptional activation and have also been implicated in several tumorigenic processes (Brown et al., 1998). Etv5(Erm) is expressed in multipotent NCCs, peripheral neurons, and satellite glia and is required in early NC development (Paratore et al., 2002a). Blocking Etv5 function in NCCs interferes with neuronal specification but not with glial specification. Neuronal progenitor survival and proliferation are not affected, but the proliferation rate of glial cells was dramatically reduced, suggesting a glia-specific role for Etv5 in controlling cell cycle progression (Paratore et al., 2002a). Etv5 distinguishes satellite glia from Schwann cells beginning early in rat PNS development. In developing DRGs, Etv5 is present in presumptive satellite glia and in neurons, but not at any developmental stage in peripheral nerve Schwann cells. Etv5 is downregulated in DRG-derived glia adopting Schwann cell traits in culture. In NC cultures, Etv5-positive progenitor cells give rise to two distinct glial subtypes: Etv5-positive, Oct-6-negative satellite glia in response to GGF2, and Etv5-negative, Oct-6-positive Schwann cells in the presence of serum and the adenylate cyclase activator forskolin. Etv5-positive NC-derived progenitor cells and presumptive satellite glia can acquire Schwann cell features (Hagedorn et al., 2000).

The expression of Etv5 and the related gene Etv4(Pea3) correlate closely with Fgf8 and Fgf3 domains, and Fgf8 is necessary for normal levels of Etv5 and Etv4 in NC. Inhibition of Fgf signaling by overexpressing Sprouty4 or an Fgf inhibitor leads to a loss of all Etv5 and Etv4 expression domains. Ectopic Fgf3 or Fgf8 results in ectopic Etv5 and Etv4 (Raible and Brand, 2001). In zebrafish, transcriptional activation by Etv4 in the NCCs is enabled by the Erk and PKA pathways (Brown et al., 1998). In *Xenopus*, Etv1(Er81) transcripts are found in NCCs, eyes, otic vesicles, and pronephros. *Etv1* transcription can also be stimulated by bFGF and eFGF in animal cap explants, and expression of dominant-negative FGF receptor in animal caps or embryos blocks *Etv1* transcription, suggesting that expression of this *Ets* gene also requires active FGF signaling (Munchberg and Steinbeisser, 1999).

4.4 Erg

Like other Ets family transcription factors, Erg appears to be involved in several fundamental developmental steps in murine embryogenesis, including EMT, cell migration, and differentiation (Vlaeminck-Guillem et al., 2000). *Xenopus Erg* encodes at least two alternatively spliced isoforms.

Transcripts are restricted to the forming endocardium, blood vessel endothelial cells, and NC-derived pharyngeal arch mesenchyme. Ectopic expression of Erg in *Xenopus* embryos causes multiple developmental defects. Dorsal injection causes a shortened anterior-to-posterior axis and severe defects in eye and somite morphogenesis, whereas ventral injection results in a posteriorization of cells and ectopic endothelial cell differentiation. In both cases, erythrocytes accumulate in structures not connected with the blood circulatory system. Erg is also implicated in cell motility and development of the circulatory system (Baltzinger et al., 1999). *Erg* gene expression predominates in mesodermal tissues, including the endothelial, precartilaginous, and urogenital areas, and also in migrating NCCs. Comparisons with *Fli1*, the gene most closely related to *Erg*, reveal overlapping expression of the two mRNAs, suggesting they may contribute to related functions. The human *Erg* gene is involved in Ewing's sarcoma, a primitive neurectodermal pediatric tumor. In chicken embryos, Erg, which acts as a transcriptional activator through a conventional consensus ETS binding site, is expressed in mesoderm-derived tissues and, to a lesser extent, in ectoderm-derived tissues. From stage E1 to E3.5, Erg expression was widely distributed in mesodermal derivatives and NC, much like Ets1, but by E6, unlike Ets1, Erg was detected in precartilaginous condensation zones and cartilaginous skeletal primordial (Dhordain et al., 1995).

4.5 Fli1

Xenopus Fli1 has extensive homology to murine and human Fli1 proteins. Transcripts appear in the early neurula, accumulate up to the tadpole stage, and are localized to regions invaded by NCCs (Meyer et al., 1993). *Xenopus* Fli1 is expressed in regions invaded by NCCs and in regions affected by important cellular migrations and/or EMTs such as endothelial cells of the heart, blood vessels, along the pronephric duct migration pathway and at the level of hypophysis. A number of genes encoding adhesion molecules or components of the ECM are potential targets of Fli1 (Meyer et al., 1995). Involvement of Fli1 in Ewing's sarcoma, a putative neurectodermal tumor, and Fli1 expression studies in mouse and *Xenopus*, suggest a role for Fli1 in NC development. In avians, Fli1 is expressed in NCCs around the time they reach their target sites. Among NC-derived lineages, Fli1 is restricted to mesenchymal cells; expression is maintained at later stages in cartilage of both NC and mesodermal origin (Mager et al., 1998).

4.6 Ets TRANSLOCATIONS

Ewing family tumors (EFTs), such as Ewing Sarcoma and other peripheral primitive neurectodermal tumors (pPNETs), are pediatric cancers displaying limited neural differentiation and of probable NC origin (Eliazer et al., 2003; Knezevich et al., 1998; Mohindra et al., 2008; Rorie et al., 2004; Staege et al., 2004). Greater than 95% of these tumors share common chromosomal translocations leading to gene fusions between the *Ewing sarcoma breakpoint region 1* (*Ewsr1*) locus and an Ets family member (*Ewsr1-Fli1* or *Ewsr1-Erg*, respectively). These fusion oncoproteins are hypothesized to function as aberrant transcription factors, either contributing to the limited neural differentiation in the tumors or inhibiting differentiation (Knezevich et al., 1998). Mis-expression of chimeric Ewsr1-Fli1 fusions proteins or wild-type Fli1 in migratory NCCs in the chicken embryo leads to significant aberrations in NC development (Coles et al., 2008). Approximately 80% of Ewing Sarcoma tumors develop in skeletal sites, but the remainder arise in almost any soft tissue location. In culture, Ewsr1-Fli1 profoundly inhibits the myogenic differentiation program of a myoblast cell line capable of differentiation into mesenchymal lineages by transcriptionally and posttranscriptionally suppressing MyoD and myogenin. EFT cell lines constitutively express CyclinD1 and have decreased expression of the cell cycle regulator p21/Cip1, even under differentiation conditions and at confluent density (Eliazer et al., 2003). Neuroblastomas (NBs), another pediatric cancer derived from the NC, display features of the SNS, but EFTs/pPNETs express markers consistent with parasympathetic differentiation. In somatic cell hybrids between NB and EFT/pPNET cells, NB-specific markers are gone, but EFT/pPNET-specific markers are unchanged, suggesting that the *Ewsr1-Fli1* fusion gene might account for loss of NB-specific markers. In support of this, heterologous expression of Ewsr1-Fli1 causes suppression of NB-specific markers and de novo expression of EFT/pPNET markers in two NB cell lines. Ewsr1-Fli1 likely contributes to the etiology of EFTs/pPNET by subverting the differentiation program of its NC precursor cell to a less differentiated and more proliferative state (Rorie et al., 2004). Genes upregulated in EFTs indicate a high similarity between EFTs and primitive NC-derived progenitor cells transitioning toward mesenchymal and/or endothelial differentiation. Ectopic expression of the EFT-specific Ewsr1-Fli1 fusion protein in HEK293 cells was not sufficient to induce the complete EFT-specific gene expression signature, suggesting that the EFT-specific gene expression profile is not merely a consequence of Ewsr1-Fli1 expression but depends on the origin of the EFT stem cell (Staege et al., 2004). Induction of Eswr1-Fli1 using a tetracycline-inducible *Ewsr1-Fli1* expression system in the rhabdomyosarcoma cell line RD results in a change in cell morphology and typical EFT features distinct from tumors formed by the parental RD cell line. Comparison of the upregulated genes with the EFT signature genes identified important Ewsr1-Fli1 downstream genes, many involved in NC differentiation. The neural phenotype of Ewing's tumors is attributable to Ewsr1-Fli1 expression and the resulting phenotype resembles the developing NC. Such tumors have a limited neural phenotype regardless of tissue of origin (Hu-Lieskovan et al., 2005). Clear cell sarcoma of tendons and aponeuroses, also referred to as malignant melanoma of soft parts, is a rare malignancy derived from NCCs. Most cases show a reciprocal cytogenetic translocation resulting in a unique chimeric fusion *Ewsr1-Atf1* gene transcript (Dim et al., 2007).

• • • •

CHAPTER 5

Fox Genes

The Fox (forkhead box) family consists of a large number of genes that contain a conserved winged helix DNA binding domain. In the mouse, there are 35 Fox genes (http://biology.pomona.edu/fox/). Fox transcription factors often are associated with control of cell differentiation, as will be demonstrated in many of the examples below. They may act as regulatory "gatekeepers" at different steps along NC differentiation pathways. Several Fox proteins have been hypothesized to act as pioneering factors, binding to DNA and perhaps required in many tissues for the recruitment of additional transcription factors and chromatin modifiers (Cuesta et al., 2007; Sekiya et al., 2009; Zaret et al., 2008).

5.1 FoxC

Foxc1 plays an important role in the NC-derived periocular mesenchyme, crucial for the development of the eye. In the anterior eye of humans and mice, NCCs populate the inner layers of the cornea, the iridocorneal angle, and the anterior part of the iris (Ittner et al., 2005; Iwao et al., 2009; Sowden, 2007). Foxc1 is expressed throughout development of the eye, and mutations in human Foxc1, along with mutations in Pitx2, have been associated with Axenfeld–Rieger anomaly (ARA), an autosomal dominant form of anterior segment dysgenesis (ASD) (Lines et al., 2002; Sowden, 2007). Upon identification of *Foxc1* as a candidate gene for ARA/ASD, mouse models were identified allowing for detailed expression and functional studies. One mouse model of ARA is the *congenital hydrocephalus* (*ch*) mutant mouse (Hong et al., 1999), which harbors a nonsense mutation in the *Foxc1* locus. Mice heterozygous for *Foxc1^{ch}* have anterior segment defects, whereas homozygous mutants have cranial NC-derived skeletal defects leading to lethal hydrocephalus among other developmental abnormalities (Hong et al., 1999).

Foxc1 and Pitx2 act together to specify primarily NC-derived mesenchymal progenitor cells as they migrate around the embryonic "optic cup." Foxc1 and Pitx2 also regulate differentiation of these same mesenchymal cells to produce the various cell types of the anterior segment (Sowden, 2007). In culture, TGFβ enhances Foxc1 and Pitx2 expression, and anterior segment morphogenesis requires TGFβ signaling (Ittner et al., 2005). TGFβ2 from the lens is required for Foxc1 and Pitx2 expression in the NC-derived cornea and angle (Ittner et al., 2005; Iwao et al., 2009). Inhibition of TGFβ signaling (Ittner et al., 2005) or reduction of TGFβ2 signaling in mouse embryonic NCCS by disruption of heparan sulfate synthesis (Iwao et al., 2009) leads to reduced phosphorylation of Smad2 and downregulation of Foxc1 and Pitx2 (Iwao et al., 2009), resulting in ARA/ASD-like ocular defects due to NCC differentiation and survival defects (Ittner et al., 2005). Posterior to the lens, inactivation of TGFβ signaling specifically in NCCs causes morphogenetic and patterning defects similar to human persistent hyperplastic primary vitreous (Ittner et al., 2005). The expression and perhaps function of Foxc1 is likely conserved in the eye; in later stages of *Xenopus* embryogenesis, *Foxc1* transcripts can be found within NCCs surrounding the eye. It is also detected in the mandibular, hyoid, and pharyngeal arches and in the heart (Koster et al., 1998).

Another FoxC gene expressed in the NC is *Foxc2*. In early mouse embryos, Foxc2 is expressed in cranial NC and can be found later in the dorsal aorta. Foxc2 heterozygous mice have ASD-like eye defects (Smith et al., 2000), and Foxc2-null mice have defects in the NC-derived cranial skeleton and in NC-mediated remodeling of the aortic arch resulting in embryonic or perinatal lethality (Iida et al., 1997). During heart development, Foxc1 and Foxc2 likely share some functional redundancy (Kume et al., 2001). Both genes are expressed in cardiac NCCs and the second heart field, endocardium, and proepicardium. Compound-null mutants have a wide spectrum of cardiac abnormalities including defects in the NC-mediated septation of the outflow tract, likely due to extensive cardiac NCC apoptosis during migration (Seo and Kume, 2006).

5.2 FoxD

Foxd3 is the most studied Fox gene in the NC and likely the most critical for NC development. Expression in premigratory and migratory NCCs was first described in the early chicken embryo (Yamagata and Noda, 1998), and soon after in the mouse embryo (Labosky and Kaestner, 1998). In all vertebrates studied, Foxd3 expression begins in the dorsal NT as commitment to NC fate occurs, and as the cells migrate to their final destinations, Foxd3 expression diminishes (Dottori et al., 2001; Hromas et al., 1999; Yamagata and Noda, 1998) (Figure 5.1). One exception is the melanoblasts, which migrate away from the dorsal neural tube at a later stage and do not express Foxd3 (Kos et al., 2001).

The only other members of the FoxD subgroup with reported NC expression or function are *Xenopus* Foxd1 and Foxd2. Foxd1 is expressed in prospective NCCs during early embryonic development but is extinguished in the NC at the time of migration. Grafting experiments and overexpression of Foxd1 suggest that it represses NCC migration (Gomez-Skarmeta et al., 1999). Foxd2 expression was observed in cranial NCCs surrounding the eye, migrating into the second and third visceral pouches, and also in ethmoidal and mandibular processes of the facial skeleton (Pohl and Knochel, 2002).

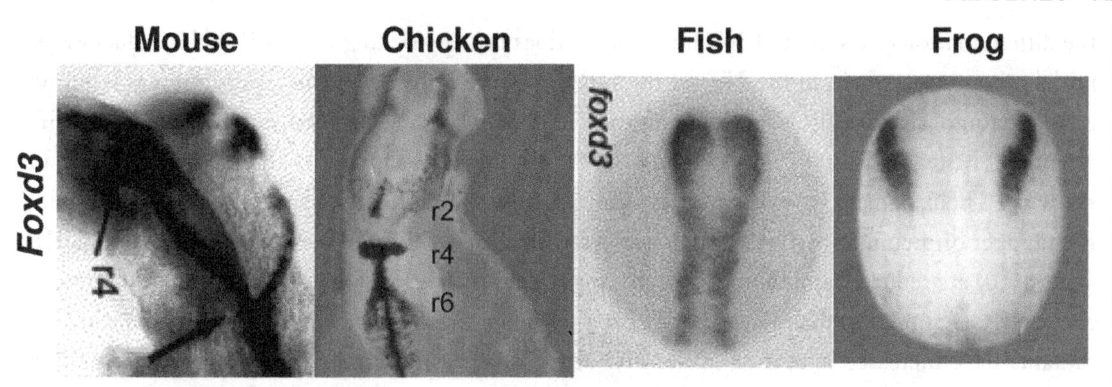

FIGURE 5.1: Whole-mount in situ hybridizations show *Foxd3* in mouse, chicken, fish, and frog embryos. Mouse: Labosky and Kaestner, 1998, 8.5-dpc embryo, dorsal view, anterior at top; chicken: 13-somite embryo, dorsal view; Theveneau et al., 2007, 13-somite-stage embryo; zebrafish: Rau et al., 2006, three-somite embryo, dorsal view; *Xenopus*: Sato et al., 2005, early neurula embryo, dorsal view.

5.2.1 Foxd3 in Neural Crest Progenitor Cells

In the chicken embryo, misexpression of Foxd3 within the dorsal NT causes an expansion of the NC domain and represses differentiation of interneurons (Dottori et al., 2001; Kos et al., 2001). Cells ectopically expressing Foxd3 upregulate Cadherin7, delaminate, and migrate away from the NT, and cells maintaining expression of Foxd3 do not differentiate normally (Dottori et al., 2001). Here, Foxd3 acts independently of Slug, another early NC marker, to promote the development of NCCs from the developing NT (Dottori et al., 2001). In the mouse embryo, Foxd3 is required for maintenance of multipotent NC progenitors. An NC-specific deletion of a floxed allele of Foxd3 results in a broad loss or severe reduction of NC derivatives, including the craniofacial skeleton, pharyngeal arch NCCs, sensory ganglia, DRGs, cranial ganglia, and enteric ganglia (Teng et al., 2008). The cardiac NC progenitor population also seems to be greatly reduced. Foxd3 likely plays a role in controlling NC survival because these Foxd3 conditional mutants have an increase in NC apoptosis (Teng et al., 2008). An additional role in early NC development for Foxd3 suggested by work in human cells is the maintenance of a progenitor state through prevention of terminal quiescence, possibly by inhibiting the cyclin-dependent kinase inhibitor p21. RA-induced expression of neuronal differentiation markers in pluripotent human teratocarcinoma cells was not influenced by forced overexpression of Foxd3, but an RA treatment-associated inhibition of proliferation, due to an increase in p21, was overcome by Foxd3 overexpression (Hromas et al., 1999).

In zebrafish, knockdown of Foxd3 results in an early reduction of several NC lineages, including jaw cartilage, sympathetic and enteric neurons, glia, and iridophores. Foxd3 is required for the differentiation of a subset of NC lineages but does not appear to greatly affect NC induction or initial migration away from the NT. In the DRGs, where Foxd3 expression persists, it is required cell-autonomously for DRG development, but expression is then reduced as the DRGs fully differentiate (Lister et al., 2006). Zebrafish *Foxd3^{sym1}* homozygous mutants that contain a mutation in the winged helix domain have defects very similar to those observed after Foxd3 knockdown. Again, the number of premigratory NCCs is not affected, but the levels of other early NC genes (Snail1b and Sox10) are reduced, and there is delayed migration and a reduction in the number of migratory trunk NCCs. Similar to the mouse, Foxd3 plays a role in the survival of some NCCs; the *Foxd3^{sym1}* mutants have increased apoptosis in the NC at the level of the hindbrain (Stewart et al., 2006). Zebrafish *Foxd3$^{mother\ superior}$* mutants that have a disruption in a distal regulatory element likewise have similar defects and, similar to the *Foxd3^{sym1}* mutants, have a reduction of the NC transcription factors Snail, Sox9b, and Sox10. In these mutants, too, the premigratory NCCs form in normal numbers, suggesting that Foxd3 maintains NC progenitor pools (Montero-Balaguer et al., 2006). Although the study of Foxd3 has been limited primarily to the typical model organisms, Foxd3 expression is observed in the NC-derived osteogenic cells of the turtle plastron bone (Cebra-Thomas et al., 2007). Little is known about Foxd3 gene regulation, but one factor identified upstream of Foxd3 is the Disrupted in schizophrenia 1 protein (Disc1). Loss of Disc1 in zebrafish embryos results in failure of the cranial NCCs to migrate away from the midline dorsal to the NT. Disc1 functions in the transcriptional repression of Foxd3 and Sox10, mediating cranial NCC migration away from the NT and cranial NCC differentiation. Continued expression of Foxd3 and Sox10 affects cranial NCC development, leading to loss of craniofacial cartilage and expansion of cranial peripheral glia. Foxd3 and Sox10 have many functions in cranial NCCs, including maintenance of precursor pools, initiation of migration, and induction of differentiation (Drerup et al., 2009). Reduction of Foxd3 or Sox10 also results in loss or reduction of neuromodulatory GnRH cells of the terminal nerve possibly arising from cranial NC (Whitlock et al., 2005).

5.2.2 NC Induction and Regulation of Foxd3

In *Xenopus*, Foxd3 is initially expressed within the Spemann organizer where it functions in mesoderm induction and then later in premigratory NCCs (Pohl and Knochel, 2001). Much of the work examining the role of Foxd3 within the complex gene regulatory network that promotes NC induction has been performed in *Xenopus*. Foxd3 is one of a complex network of factors including Msx1, Zic1, Zic5, Hes4, Snail, Slug, Pax3, Pax7, Sox10, Sox9, and Twist1 that act to specify the NC. Foxd3 expression is downstream of Msx1, Hes4, Zic1, Pax3, Snail, Sox10, and Sox9 (Aybar et al., 2003; Honore et al., 2003; Nagatomo and Hashimoto, 2007; Osorio et al., 2009; Sato et al., 2005; Tribulo et al., 2003; Yan et al., 2005). However, these networks also include feedback mechanisms, as Foxd3 also regulates expression of several genes initially upstream of Foxd3, such as Sox10

(Cheung et al., 2005; Dutton et al., 2008). Also, many of these genes regulate expression of the other genes in the network once they are activated, and this is true for Foxd3. Chicken LSox5 is initially expressed in premigratory and then migratory NC after FoxD3 expression. However, forced expression of LSox5 enlarges the NC domain and prolongs NC segregation, resulting in overproduction of Foxd3-expressing cranial NCCs, another example of a feedback mechanism (Perez-Alcala et al., 2004). Early expression of Foxd3 in the prospective NC also requires activation by the Wnt pathway and is inhibited by BMP signaling (Pohl and Knochel, 2001; Taneyhill and Bronner-Fraser, 2005). In *Xenopus*, Foxd3 expression in the prospective NC is similar to expression of other early NC markers such as Slug and Zic1 and is itself required for NC induction, at least in part through regulation of Slug (Sasai et al., 2001). Overexpression of Foxd3 in the embryo or ectodermal explants results in expanded expression of NC markers. A dominant-negative Foxd3 construct inhibits NC differentiation in the embryo, but this can be rescued by coinjection of Slug. In animal cap explants, dominant-negative Foxd3 can inhibit NC differentiation even in the presence of the NC-inducers Slug and Wnt3a, and Foxd3 is necessary for the induction of Slug by Zic proteins (Sasai et al., 2001). Expression of Snail, Slug, and Foxd3 together leads to delamination from the neural tube (Tucker, 2004). Upon delamination, Foxd3 becomes downregulated in most lineages at the start of migration (Taneyhill and Bronner-Fraser, 2005).

In addition to *Slug*, there are a few genes that are known to lie downstream of Foxd3, but not many direct targets of this winged helix transcription factor are known. Foxd3 regulates expression of cell adhesion molecules N-cadherin, Integrinβ1, Laminin, and Cadherin7, required for NC migration (Cheung et al., 2005). In *Xenopus* mesoderm induction, Foxd3 acts as a repressor (Steiner et al., 2006; Yaklichkin et al., 2007). Foxd3 recruitment of Groucho corepressors is essential for the transcriptional repression of target genes and induction of mesoderm in *Xenopus* (Yaklichkin et al., 2007). Recent evidence in embryonic stem cells (ESCs) supports the hypothesis that Foxd3 acts as a pioneer factor to open chromatin and affect downstream gene expression. Foxd3 regulates demethylation at a CpG in the Alb1 enhancer, implicating this stem cell protein in maintenance and/or generation of transcriptional competence in ESCs (Xu et al., 2009). This has not yet been investigated in NC or NCSCs.

5.2.3 Foxd3 Suppresses the Melanocyte Lineage

One well-characterized target of Foxd3 is *Mitf*. The first NCCs that migrate out from the NT are specified as neurons and glia. The subsequent wave of migrating NCCs will give rise to melanocyte precursors (Kos et al., 2001; Thomas and Erickson, 2009). Mitfa, a master regulator of melanogenesis, is only expressed in Foxd3-negative NCCs, and the number of cells expressing Mitfa increases in Foxd3 mutants. Foxd3 controls the lineage choice between neurons and glia versus melanocytes by sequestering Pax3 to indirectly repress Mitf expression during the early phase of NC migration.

Foxd3 thus represses melanogenesis and is expressed exclusively in neural and glial precursors, whereas Mitf is expressed only in melanoblasts. Ectopic expression of Foxd3 represses Mitf in cultured NCCs and in melanoma cells (Thomas and Erickson, 2009). There is also evidence that Foxd3 can act directly on the Mitfa promoter to repress its expression (Curran et al., 2009; Ignatius et al., 2008). Misexpression of Foxd3 in late-migrating NCCs suppresses development of the melanocyte lineage, whereas knockdown of Foxd3, both in vivo and in vitro, results in expansion of the melanoblast population and perhaps loss of neuronal and glial precursors (Kos et al., 2001). In zebrafish, Hdac1 is required to suppress Foxd3 expression in a subset of the NC, thus de-repressing Mitfa and allowing melanogenesis to proceed (Ignatius et al., 2008). In zebrafish *Hdac1^{colgate}* mutants, Foxd3 expression is increased and prolonged, and although normal numbers of premigratory NCCs are induced, fewer melanoblasts are specified, and there is a delay in differentiation and decreased migration of melanophores and melanophore precursors. Mitfa expression and melanophore defects in *Hdac1^{colgate}* mutants are rescued by partial reduction of Foxd3 expression.

5.3 FoxE

Mutations in human Foxe1 are linked to the Bamforth–Lazarus syndrome, characterized by hypothyroidism and cleft palate (Nakada et al., 2009). Zebrafish Foxe1 is expressed in the thyroid, pharynx, and pharyngeal skeleton during development (Nakada et al., 2009). Morpholino knockdown of Foxe1 function resulted in disrupted craniofacial and pharyngeal skeleton development, likely as a result of suppressed chondrocytic proliferation. The initial steps of NC migration and pharyngeal arch specification occur normally, but later chondrocyte proliferation and differentiation was perturbed, as indicated by a reduction in Sox9a, Col2a1, and Runx2b (Nakada et al., 2009).

5.4 FoxF

In *Xenopus*, Foxf1a is expressed in cranial NC (Koster et al., 1999). In mouse, the cranial NC expression of both Foxf1 and Foxf2 is restricted to the palatal mesenchyme, where their expression levels, along with those of Osr2, are regulated by Shh-Smo signaling (Lan and Jiang, 2009). Foxf2 expression in the mesenchyme around the oral cavity is stronger than Foxf1 expression, and Foxf2 null mice have a cleft palate, the only reported defect (Ormestad et al., 2004). Foxf1-null mutants exhibit early lethality due to extraembryonic- and lateral plate mesoderm-related defects (Ormestad et al., 2004).

5.5 FoxJ

In the mouse, Foxj3 is expressed in the neuroectoderm at early stages (starting at E8.5) and continues in the NC and NC-derived structures, such as the facioacoustic, trigeminal, and DRGs (Landgren and Carlsson, 2004).

5.6 FoxN

Xenopus Foxn2 is detected in the early eye field and later in the pharyngeal arches, vagal ganglion, and developing retina (Schuff et al., 2006). *Xenopus* Foxn3 is expressed in NCCs and the early eye field and in pharyngeal arches (Schuff et al., 2006; Schuff et al., 2007). Knockdown of Foxn3 results in defects of the cranial NC-derived jaw cartilage, the cranial nerves, and the eye (Schuff et al., 2007). In the developing jaw, NC migration occurs normally, but NC differentiation is disrupted. In the eye, loss of Foxn3 results in increased apoptosis. Foxn3 may regulate craniofacial and eye development in part by the recruitment of two HDAC proteins Sin3 and Rpd3 that were identified as potential binding partners of Foxn3 (Schuff et al., 2007).

5.7 FoxO

Genes of the FoxO subgroup are often important for cell cycle regulation. Both Foxo1 and Foxo3a have been demonstrated to play a role in rat NC-derived enteric nervous system precursor survival and neurite extension, acting as substrates of Akt in a PI3K-dependent pathway (Srinivasan et al., 2005). The Gdnf protein in ENS precursors induces phosphorylation of Akt leading to a decrease of Foxo1 and Foxo3a. Misexpression of active Foxo1 induces ENS precursor death, whereas a dominant negative represses ENS precursor cell death (Srinivasan et al., 2005). In *Xenopus*, Foxo1 is expressed in head mesenchyme anterior to the eye and within the pharyngeal arches. Foxo3 is expressed in NCCs at the late neurula stage and later in the head and pharyngeal arches (Pohl et al., 2004).

5.8 FoxS

The murine Foxs1 gene is expressed in NC-derived cells such as cranial ganglia, DRGs, and neurons of the enteric ganglia, as revealed by a knock-in allele placing the *lacZ* gene into the Foxs1 locus (Heglind et al., 2005). Foxs1 mutant mice have affected motor function and body weight, but this may be due solely to the function of Foxs1 in the CNS (Heglind et al., 2005). Foxs1 specifically marks early sensory neurons and sensory neuron precursors of the trunk, and is expressed exclusively in both NC-derived and ectodermal placode-derived peripheral sensory neurons, but not in non-neuronal NC-derived cell types (Montelius et al., 2007). Expression of Foxs1 does not overlap with Sox10 expression, suggesting that acquisition of Foxs1 expression represents an important lineage restriction in the differentiation of multipotent NCCs toward a sensory neuron fate. Migrating NC-derived sensory neuron precursors that express Foxs1 form clusters in the developing ganglion and exhibit reduced proliferation compared to surrounding Sox10-positive cells, and begin to express Ngn1 and Brn3 as the DRG condenses (Montelius et al., 2007).

• • • •

CHAPTER 6
Homeobox Genes

6.1 Alx

The aristaless-related homeobox (Alx) family consists of four members in mammals: Alx1, Alx3, Alx4, and Arx. The Alx paired-type homeodomain binds DNA at specific sites either as a homodimer or as a heterodimer with other paired-type homeodomain factors. Both Alx1 and Alx4 are expressed in the NC-derived first pharyngeal arch and craniofacial mesenchyme and their derivatives and in limb bud mesenchyme. In vitro, Alx1 and Alx4 can form heterodimers and can each activate transcription of reporter genes in a similar manner, suggesting functional redundancy. *Alx1*-null; *Alx4*-null compound mutants demonstrate a genetic interaction indicating a role for both genes in nasal cartilage fusion, mandible patterning, and other aspects of craniofacial development (Qu et al., 1999). Similar redundancy is seen with Alx3 and Alx4. *Alx3*-null mice are seemingly normal, but compound *Alx3*-null; *Alx4*-null mutants have severe craniofacial defects not observed in *Alx4*-null single mutants (Beverdam et al., 2001). These defects include cleft nasal regions and malformation or loss of facial bones and other NC-derived skull elements. In these *Alx3*-null; *Alx4*-null compound mutants, an increase in apoptosis in the outgrowing frontonasal process is the likely cause of the craniofacial defects (Beverdam et al., 2001).

6.2 Barx1

The homeobox transcription factor Barx1 (with some homology to *Drosophila* Bar, expressed in the eye and other sensory organs) was identified in the mouse as a protein that binds to a regulatory element of the *Ncam1* promoter. Barx1 is strongly expressed in parts of the head and neck mesenchyme, especially in parts of the first and second pharyngeal arches where it is generally restricted to NC-derived tissues, including mesenchyme associated with the olfactory epithelium, the primary and secondary palate, the stroma of the submandibular gland, and molar papillae. By E16.5, Barx1 expression in the head is only detected in the developing molars and may be involved in delineating molar identity from developing incisor identity (Tissier-Seta et al., 1995). Human Barx1 is expressed in testis, heart, iris, craniofacial mesenchyme of NC origin, and developing teeth (Gould and Walter, 2000). During development of the pharyngeal arch derivatives, Barx1 is downstream of members of the Dlx gene family. In *Dlx1*-null; *Dlx2*-null compound mutant embryos, Barx1 expression is lost in favor of the chondrogenic marker Sox9 (Thomas et al., 1997). Loss of Dlx2a in zebrafish causes ectopic Barx1 expression in the dorsal ceratohyal arch (Sperber et al., 2008). In zebrafish, Barx1 is expressed in the rhombencephalic NC and in NC-derived pharyngeal arch mesenchyme, and this expression can be induced by BMP4 and is maintained through FGF signaling. Barx1 expression is necessary for proliferation of osteochondrogenic progenitors. Knockdown of Barx1 results in reduced and malformed cartilage elements due to reductions in chondrocyte differentiation and condensation in conjunction with loss of osteochondrogenic markers Col2a1, Runx2a and Chondromodulin, and the odontogenic marker Dlx2 (Sperber and Dawid, 2008). Like many genes expressed in the NC-derived pharyngeal arch mesenchyme, Barx1 is also regulated by Endothelin signaling and its expression is significantly reduced in *Edn1*-null embryos (Clouthier et al., 2000).

6.3 Dlx1, Dlx2, Dlx3, Dlx5, Dlx6

The distalless homeobox (*Dlx*) gene family has six homologs in mammals. The six mouse Dlx genes are primarily expressed in the developing forebrain and cranial NC derivatives, particularly during NC migration, patterning of the orofacial skeleton, and tooth initiation and development (Berdal et al., 2000; Davideau et al., 1999; McGuinness et al., 1996; Merlo et al., 2000; Weiss et al., 1998). The mammalian *Dlx* genes are expressed at various times during tooth development in a manner related to their genomic organization, much like an "odontogenic Hox" code (Thomas et al., 1997; Weiss et al., 1998). Null alleles of *Dlx1*, *Dlx2*, *Dlx3*, and *Dlx5* in the mouse have revealed functions in craniofacial patterning, sensory organ morphogenesis, osteogenesis, and placental formation (Merlo et al., 2000). However, there is a substantial degree of functional redundancy shared between members of the Dlx family, making functional analysis of these genes more difficult without using compound-null mutants (Kraus and Lufkin, 2006). In zebrafish, loss of *Dlx* genes correlates with loss or abnormal morphology of craniofacial cartilage elements except for those that originate from NCCs of the midbrain region and do not express *Dlx* genes. Similar to *Hox* genes, the expression of zebrafish *Dlx* genes in NCCs migrating from the hindbrain and in visceral arch primordia has also been shown to be sensitive to treatment with RA (Ellies et al., 1997).

6.4 Dlx1 AND Dlx2

Dlx2 is required for differentiation of subsets of cranial NC and forebrain cells (McGuinness et al., 1996) and is particularly important in the NC-derived pharyngeal arch mesenchyme. Dlx1 and Dlx2 pattern the teeth by specifying a subpopulation of cranial NC as odontogenic, allowing for molar development. In mice carrying null mutations for both *Dlx1* and *Dlx2*, the NC-derived ectomesenchyme underlying the maxillary molar epithelium loses its odontogenic potential, changing fate from odontogenic to chondrogenic. These mice do not develop maxillary molars, but the incisors and mandibular molars are normal (Thomas et al., 1997). Dlx2 is also expressed in the mandible and maxilla, and is induced by signals from the mandibular and maxillary arch epithelia (Ferguson et al., 2000). Dlx1 expression extends to the third pharyngeal arch, which contributes to the thymus. *Dlx1*-null mice do not have any obvious thymus developmental defects, but there may be functional redundancy shared by other Dlx family members, particularly Dlx2. Transcripts for multiple Dlx family members were detected in a screen of an E13.5 thymus cDNA library, and *Dlx1* and *Dlx2* were detected in adult murine thymus and Thy1-positive thymocytes (Woodside et al., 2004).

6.5 Dlx3

In chicken embryos, Dlx3 is expressed in the distal portion of the first and second pharyngeal arch mesenchyme (Pera and Kessel, 1999). In mice, Dlx3 expression in the pharyngeal arches is dependent on Endothelin signaling and is downregulated in mice deficient for downstream effectors of Edn1, Gα(q), and Gα(11) (Ivey et al., 2003). Studies of Dlx3 during newt regeneration indicate that it is expressed in cells with the ability to contribute to ventral root ganglia and spinal ganglia, but only during regeneration, and not normal development. These data suggest that Dlx3 is present in cells with NC-like properties and the potential for repair (Nicolas et al., 1996).

6.6 Dlx4

A role for Dlx4 in the NC has not been reported.

6.7 Dlx5

In the mouse, early Dlx5 expression becomes localized to the anterior neural ridge, defining the rostral boundary of the neural plate, and extends caudolaterally, marking the prospective NC (Yang et al., 1998). Dlx5 is later expressed in a BMP-dependent manner in the developing skull and mandibular bones but not in the maxilla (Ferguson et al., 2000; Holleville et al., 2007). Like Dlx2, Dlx5 mandibular expression is induced by signals from the mandibular and maxillary arch epithelia (Ferguson et al., 2000). Dlx5 is also involved in initial steps of membranous differentiation of the calvaria and can induce Runx2 and osteoblast differentiation in cultured embryonic suture mesenchyme (Holleville et al., 2007). In humans, Dlx5 is primarily detected first in the mandible and then later in the maxilla, and finally is restricted to progenitor cells of the developing teeth and bones and cartilages of the mandible and maxilla. In the developing teeth, human Dlx5 is expressed in NC-derived mesenchyme and in a subset of dental epithelia (Davideau et al., 1999).

6.8 Dlx6

Dlx6 is another *Dlx* gene regulated by Endothelin1 signaling. Edn1 from neighboring cells acts on cranial NC-derived ectomesenchymal cells expressing Ednra to regulate expression of crucial genes such as Dlx6 and the Dlx6 downstream target Hand2 (Fukuhara et al., 2004; Ivey et al., 2003).

Endothelin signaling is critical for Dlx6 and Hand2 expression in the mandibular arch mesenchyme in a short time window between approximately E8.75 and E9.0 until E9.5. Later, Dlx6 and Hand2 expression is maintained or regulated by FGF and other signals from the epithelium (Fukuhara et al., 2004).

6.9 Emx

The *empty spiracles*-related homeobox genes Emx1 and Emx2 are expressed in the forebrain and are essential for forebrain development (Chiba et al., 2005; Williams et al., 1997). In NT and NC-like cells derived from ES cells by RA treatment, Emx1 and Emx2 expression occurred in response to Noggin treatment, which directs NT-like structures toward a forebrain fate (Chiba et al., 2005). Overexpression of Emx1 and Emx2 in NC-derived melanocytes downregulates the melanocyte-specific differentiation genes Mitf, Tyrp1, Dct and Tyr, and constitutive expression of Emx genes alters pigment cell morphology and growth properties. Emx regulation of Mitf suggests that one normal function of Emx genes is to inhibit Mitf activation in nonmelanocyte neuroepithelial derivatives (Bordogna et al., 2005).

6.10 Gsx

The goosecoid-related homeobox gene family includes its namesake, Goosecoid (Gsc), and the related factor Gsc2. In zebrafish, Gsc exhibits two independent phases of expression: early, in cells anterior to the presumptive notochord, and later, in NC derivatives in the larval head (Schulte-Merker et al., 1994). Zebrafish Gsc expression in the pharyngeal arches is dependent on Endothelin signaling, Mef2c function, and Dlx2 expression (Miller et al., 2007; Sperber et al., 2008). Pharyngeal arch expression of Gsc is also absent in *Edn1*-deficient and *Ednra*-deficient mouse embryos, suggesting it is one of the downstream signals triggered by activation of Ednra (Clouthier et al., 1998; Clouthier et al., 2000). In mammals, Gsc marks caudal NC-derived mesenchymal cells of the mandible, which do not receive Fgf8 signals from the rostral epithelium and give rise to the distal part of the lower jaw, whereas rostral NC mesenchyme does receive Fgf8 and gives rise to odontogenic cells. As in zebrafish, Endothelin signaling also helps to maintain positional information (Tucker et al., 1999). Gsc is also involved in patterning and morphogenesis of the NC-derived mesenchyme of the middle ear, a pharyngeal arch derivative, and Gsc is critical for tympanic ring development (Mallo, 2001). In humans, the other member of the Gsx family, Gsc2, is in the region within 22q11 that is deleted most consistently in patients with DiGeorge Syndrome/VCFS, suggesting a possible link to NC development (Gottlieb et al., 1998).

6.11 Hlx

In mouse, the H2.0-like homeobox (Hlx) gene is expressed in intestinal and hepatic mesenchyme. Hlx is required for embryonic intestine and liver growth, and homozygous loss of Hlx causes embryonic lethality. Hlx function is indirectly connected to NC function in ENS development. Without Hlx, the development of the ENS, which requires interaction between the migrating NCCs and the gastrointestinal tract mesenchyme, is abnormal. In these mutant embryos, NCCs do not enter the intestine and are primarily restricted to the lateral stomach mesenchyme (Bates et al., 2006).

6.12 Hmx1

Hmx genes comprise a novel gene family. In mouse, there are three *Hmx* genes: *Hmx1*, *Hmx2*, and *Hmx3*. Hmx2 and Hmx3 are expressed in the CNS, but Hmx1 is divergent and is expressed in NC-derived DRGs, sympathetic ganglia, and vagal nerve ganglia. In addition, Hmx2 and Hmx3 are expressed in the otic vesicle, and Hmx1 is strongly expressed in the developing eye (Wang et al., 2000).

6.13 Irx

In *Xenopus*, the Iroquois-related homeobox gene Iro1 is one of the genes induced in the neural plate border, which includes placodal and NCC precursors, and induction is controlled by a precise level of BMP activity needed during signaling from the epidermis to the neural plate (Glavic et al., 2004a; Mayor and Aybar, 2001). Iro1, Notch, and the Notch target gene *Hes4* are all expressed in the prospective NC territory, whereas Notch ligands Delta1 and Serrate are expressed in cells surrounding the prospective NCCs. An activator fusion form of Iro1 results in enlargement of the NC territory, whereas blocking Iro1 activity inhibited NC marker expression. Activation of Iro1 and Notch signaling resulted in upregulation of *Hes4* and inhibition of *Bmp4* transcription during NC specification. Iro1 lies upstream of the cascade regulating *Delta1* transcription. During early gastrulation, Iro1, as a positive regulator, and Snail as a repressor, act to restrict Delta1 expression at the border of the prospective NC territory. Additional signals then induce the production of NCCs (Glavic et al., 2004b).

There are six iroquois homeodomain (Irx) transcription factors in the mouse (Mummenhoff et al., 2001). Irx family members are implicated in a variety of early developmental processes, including neural prepatterning, tissue differentiation, NC development, and cranial placode formation (Feijoo et al., 2009). An additional Irx family member, Iro7, is found in zebrafish. During zebrafish gastrulation, Iro7 and Iro1 are expressed in a region of the dorsal ectoderm that includes the prospective midbrain–hindbrain domain, adjacent NC, and the trigeminal placodes in the epidermis (Itoh et al., 2002; Lecaudey et al., 2001). Early expansion of Iro1 and Iro7 expression in *headless* and *masterblind* Wnt signaling mutants correlates with expansion of the midbrain–hindbrain boundary domain, the NC, and trigeminal neurons. Knockdown of Iro7 alone shows it is essential for specification of neurons in the trigeminal placode, whereas knockdown of Iro1 and Iro7 together uncovers essential roles in NC development, where they may function as transcriptional repressors (Itoh et al., 2002).

6.14 Lbx1 AND Lbx2

The *ladybird homeobox* (*Lbx*) gene family has two members in the mouse, *Lbx1* and *Lbx2*. Lbx1 is found in hypaxial musculature, developing dorsal spinal cord neurons, and within a cardiac NC subpopulation required, along with myocardial cell proliferation and differentiation, during tubular heart formation (Schafer et al., 2003). *Lbx1*-null mice have heart looping defects, increased myocardial proliferation, and changes in gene expression, but no defects in NCC migration, suggesting that Lbx1 is needed for specification of this NC subpopulation. The *Lbx1* promoter driving a *lacZ* reporter demonstrates that Lbx1 promoter activity is upregulated in hearts of *Lbx1*-heterozygous, $Pax3^{Splotch1H/Splotch1H}$ compound mutant embryos and *Lbx1*-null mice, indicating that Pax3 and Lbx1 participate in a negative regulatory feedback (Schafer et al., 2003). Lbx2 is expressed in the developing urogenital and nervous systems. Lbx2-null mice appear healthy and fertile. *Lbx2-lacZ* null mice intercrossed with $Pax3^{Splotch}$ heterozygous mutant mice results in embryos with reduced Lbx2 expression in DRGs and cranial nerve ganglia, but not in the genital tubercle, suggesting that Pax3 is required for Lbx2 expression in affected NC-derived tissues (Wei et al., 2007).

6.15 Msx1 AND Msx2

The muscle segment-related homeobox (Msx) gene family consists of Msx1, Msx2, and Msx3 in mammals, but expression of only Msx1 and Msx2 has been described in the NC. Msx genes are expressed in a range of vertebrate-specific tissues, including NC, cranial sensory placodes, bone, and teeth (Davidson, 1995). In the mouse, Msx1 and Msx2 are expressed during critical stages of NT, NC, and craniofacial development. Msx1 knockdown in whole embryo culture during early stages of neurulation causes reduced growth of the maxillary, mandibular, and frontonasal prominences (presumably due to mesenchymal deficiencies), as well as eye, somite, and NT abnormalities; Msx2 knockdown produces similar results. Double knockdown does not cause a more severe phenotype, suggesting that there is no redundancy or synergy between the two factors (Foerst-Potts and Sadler, 1997). An *Msx1*-null allele was generated by insertion of an *nuclear lacZ* reporter gene into the *Msx1* coding region. LacZ expression from this allele in heterozygous embryos showed novel Msx1 expression in migrating NCCs. Homozygous null mice die at birth with craniofacial defects. In most regions of the face affected in these mutants, there is no overlapping expression of Msx2 (Houzelstein et al., 1997). Msx1 and Msx2 are both expressed at sites of cellular proliferation and programmed cell death, including the cranial NC. Bmp4 can regulate cell death at these sites and induce Msx1 and Msx2 expression, and Msx2 is a key regulator of apoptosis in a BMP-mediated pathway. Constitutive ectopic Msx2 expression in P19 cells results in a marked increase in aggregation-induced apoptosis but has no effect when cells are grown as a monolayer; in this system, addition of Bmp4 induces programmed cell death through Msx2 (Marazzi et al., 1997).

6.16 Pitx2

Heterozygous *Pitx2* mutations cause ocular anterior segment defects and early-onset glaucoma. Pitx2 is expressed in NC- and mesoderm-derived precursors of the periocular mesenchyme. *Pitx2* homozygous loss of function in mice results in severe disruption of periocular mesenchyme structures. Pitx2 is required specifically in the NC for specification of corneal endothelium, corneal stroma and sclera, and for normal development of ocular blood vessels (Evans and Gage, 2005). In the cardiac NC, Pitx2 may function as a target of canonical Wnt signaling, but it is not essential for cardiac NC development (Ai et al., 2006).

6.17 Prrx1 AND Prrx2

The paired-related homeobox genes *Prrx1* and *Prrx2* both play a role in epithelial–mesenchymal interactions in the pharyngeal arches. Prrx1 is expressed in the mesenchyme of facial, limb, and vertebral skeletal precursors during mouse embryogenesis and may regulate epithelial–mesenchymal interactions required for skeletal morphogenesis. *Prrx1*-homozygous-null mutants die soon after birth with defects of limb, vertebral, and NC-derived craniofacial skeletal structures caused by cellular defects in formation and expansion of chondrogenic and osteogenic precursors from undifferentiated mesenchyme (Martin et al., 1995). *Prrx1* is one gene involved in patterning and morphogenetic processes in the NC-derived mesenchyme in the developing middle ear, and it is essential for tympanic ring development (Mallo, 2001). Prrx2 is also expressed in pharyngeal arch mesenchyme (de Jong and Meijlink, 1993) and at sites of epithelial–mesenchymal interactions, including within the cranofacial mesenchyme. During tooth development from the early bud stage to the late bell stage, Prrx2 is expressed exclusively in NC-derived ectomesenchyme and derivatives. In both first molar and incisor primordia, Prrx2 expression is highest at the late cap and early bud stages and declines at the mid-bell stage. Prrx2 is also present in first and second molar primordia cultured from E13 jaw explants (Karg et al., 1997). In *Hand1; Hand2* compound mutants, Prrx2 is dysregulated in the distal craniofacial mesenchyme (Barbosa et al., 2007).

6.18 Runx1 AND Runx2

The runt homeobox gene family has three main homologs, *Runx1*, *Runx2*, and *Runx3*. In chicken and mouse, Runx1 is selectively expressed in NC-derived TrkA-positive sensory neurons to mediate activation of TrkA in migratory NCCs and promote axonal growth. Without Runx activity, TrkA expression is lost, leading to neuronal death. Overexpression of Runx1 is not compatible with migratory NC multipotency but does not induce expression of pan-neuronal genes on its own. Runx1-induced neuronal differentiation toward a TrkA-positive nociceptive sensory neuron fate depends on an existing Ngn2 proneural gene program (Marmigere et al., 2006). Forced expression of Runx1 in Sox10-expressing boundary cap NCSCs strongly increased survival and was sufficient to guide differentiation of boundary cap NCSCs toward a nonpeptidergic nociceptive sensory neuron fate (Aldskogius et al., 2009).

Runx2 is a master regulator of bone formation, and *Runx2*-null mice have a severe loss of the osteogenic skeleton. Runx2 is also expressed in the pre-hypertrophic cartilaginous skeleton of the mouse and chicken, but its function here is not clear. Unlike mouse and chicken, zebrafish and *Xenopus* require Runx2 function for early cartilage differentiation, and Runx2 is expressed in mesenchymal precursors of the cartilaginous skull (Deng et al., 2008; Kerney et al., 2007). Runx2, as an osteoblast marker, is ectopically expressed in *Sox9*-null mouse embryos, where nasal cartilages should normally be, suggesting that without Sox9, cranial NCCs may lose chondrogenic potential and adopt an osteogenic fate (Akiyama et al., 2005; Mori-Akiyama et al., 2003). Runx2a in zebrafish is also regulated by Sox9 (Yan et al., 2005), and in the absence of Fgfrl1a, both Sox9a and Runx2b are lost from the mesenchymal condensations of the pharyngeal arches (Hall et al., 2006). A number of proteins are upstream of Runx2 in the pharyngeal arch mesenchyme: Msx, Dlx5, Barx1, and Foxe1 are all positive upstream regulators of Runx2, whereas Hand2 is a negative regulator of Runx2. *Msx* genes are critical for Runx2 expression in frontonasal NCCs and differentiation of the osteogenic lineage (Han et al., 2007). In the chicken embryo, Dlx5 upregulates Runx2 and subsequent osteoblast differentiation in cultured embryonic suture mesenchyme. A dominant-negative Dlx interferes with the ability of the BMP pathway to activate Runx2 expression (Holleville et al., 2007). Foxe1 knockdown morphants, with craniofacial defects, have severe reduction in expression of Sox9a, Col2a1, and Runx2b (Nakada et al., 2009). Runx2a expression in zebrafish pharyngeal arches is lost upon knockdown of Barx1 (Sperber and Dawid, 2008). Hand2 and Runx2 are partially colocalized in the mandibular primordium of the pharyngeal arch, and downregulation of Hand2 precedes Runx2-driven osteoblast differentiation. *Hand2* hypomorphic mutant mice have upregulated and ectopic expression of Runx2 in the mandibular arch (Funato et al., 2009).

6.19 Tlx

The human T-cell leukemia homeobox gene *Tlx1* is one of the genes involved in a chromosomal translocation occurring in 5–10% of patients with T-cell acute lymphoblastic leukemia. Accordingly, murine Tlx1 is expressed in the developing spleen and *Tlx1*-null mice are asplenic (Lichty et al., 1995), but Tlx1 is also expressed in cranial NC derivatives. Two additional members of the Tlx family, Tlx2 and Tlx3, are necessary for proper ANS development (Bachetti et al., 2005). In the chicken embryo, both Tlx1 and Tlx3 are expressed in placode-derived components of the cranial sensory ganglia, but only Tlx3 is expressed in NC-derived DRG and sympathetic ganglia (Logan et al., 1998). Tlx2 is expressed in the primarily vagal NC-derived ENS, and *Tlx2*-null mice have ENS defects resembling human intestinal neuronal dysplasia type B (Puri and Shinkai, 2004). In general,

Tlx2, together with Phox2b, which is also downstream of BMP signaling and has an expression pattern similar to Tlx2, is critical for the development of NC-derived cells adopting ANS fates. Phox2b can bind a cell-specific enhancer in the 5' regulatory region of *Tlx2* and activates *Tlx2* in neuroblastoma cells. Tlx2 is upregulated by overexpression of Phox2b, and Phox2b protein carrying a mutation responsible for congenital central hypoventilation syndrome (CCHS) development has a severely hindered ability to activate *Tlx2* expression in vitro and in vivo (Borghini et al., 2006). Tlx2 enhances the activity of the Ret promoter in a neuroblastoma cell line (Bachetti et al., 2005). Tlx proteins may be involved in differentiation and maintenance of specific neuronal populations (Logan et al., 1998).

• • • •

CHAPTER 7
Hox Genes

Hox transcription factors are a conserved family of homeobox-containing proteins related to *Drosophila* homeotic Hom-C proteins. Across metazoan species, *Hox* genes are important for anterior-to-posterior patterning, and the organization of *Hox* genes within the genome correlates with their anterior-to-posterior expression within the body plan. In vertebrates, tight regulation of Hox expression is essential for specification of hindbrain segments and NC derivatives contributing to cranial ganglia and pharyngeal arches (Nonchev et al., 1996; Wilkinson, 1993). Vertebrate *Hox* genes pattern the hindbrain and pharyngeal regions of the developing head up to and including structures derived from the second pharyngeal arch. The first pharyngeal arch and more rostral regions of the head are patterned by groups of homeobox genes more diverged from the original *Hox* clusters (Cobourne, 2000). Patterning of NC derivatives at more posterior levels, such as the ENS, also requires Hox proteins to establish anterior-to-posterior patterning. There is a large amount of interplay between the various Hox factors; many regulate transcription of other *Hox* genes, and these interactions help establish distinct segmented regions, each with unique Hox expression.

7.1 ANTERIOR Hox GENES (Hox1, Hox2, AND Hox3 PARALOGOUS GROUPS)

The expression domains of the *Hox1*, *Hox2*, and *Hox3* paralogous group genes roughly correspond to the sites of origin of cranial NCCs. These genes are sensitive to RA signaling and interact with each other to establish boundaries resulting in correctly segmented rhombomeres and pharyngeal arches and appropriately patterned NCCs.

7.1.1 Hoxa1, Hoxb1, and Hoxd1

There are three vertebrate *Hox1* genes: *Hoxa1*, *Hoxb1* and *Hoxd1*. In *Xenopus*, these three genes have broadly overlapping expression patterns encompassing the prospective hindbrain and associated NCCs. Hoxa1, Hoxb1, and Hoxa2 function together with Krox20 and Mafb to pattern the hindbrain and associated NCCs (Barrow et al., 2000). Hoxa1 activity establishes the anterior boundary of Hoxb1 expression between r3 and r4; without Hoxa1, Hoxb1 is not expressed to this boundary, resulting in misexpression of other genes and subsequent misspecification of r2 through r5

(Barrow et al., 2000). Hoxa1 and Hoxb1are involved in distinct aspects of hindbrain segmentation and specification (Gavalas et al., 1998), but strong genetic interactions between Hoxa1 and Hoxb1 indicate overlapping and synergistic roles in patterning the hindbrain and cranial NCCs (Gavalas et al., 1998; Gavalas et al., 2001). Hoxa1 also regulates the expression of Cdh6, an important adhesion molecule sharing an overlapping expression with Hoxa1 in the early posterior hindbrain. Cdh6 is also expressed in NCCs migrating out of the NT to the peripheral nerves (Inoue et al., 1997). After activation in the neuroectoderm, Hoxb1 expression is restricted to r4 (Morriss-Kay et al., 1991; Studer et al., 1994). In both chicken and mouse embryos, a *Hoxb1* enhancer element drives reporter expression from r3 to r5, and a repressor element containing an RA response element restricts Hoxb1 to r4 (Studer et al., 1994). Simultaneous knockdown of all three *Xenopus Hox1* genes results in hindbrain patterning defects more severe than in single or double knockdowns, demonstrating that these genes share some functional redundancy (McNulty et al., 2005). Without the *Hox1* genes, cranial NCCs do not migrate into the pharyngeal arches, causing defects in derivatives of the pharyngeal arches such as loss of gill cartilages (McNulty et al., 2005).

In addition to regulating other *Hox1* genes, Hox1 proteins regulate other paralogous *Hox* groups; in *Xenopus*, *Hox1* genes are required for hindbrain expression of *Hox2*, *Hox3*, and *Hox4* genes (McNulty et al., 2005). Hoxa1 and Hoxa2 together play a role in patterning and morphogenesis of NC-derived pharyngeal arch mesenchyme, including mesenchyme contributing to the middle ear (Mallo, 2001). In the mouse embryo, loss of Hoxa1 or Hoxa2 alone affects multiple NC-derived structures, but *Hoxa1; Hoxa2* compound mutants embryos have more severe craniofacial defects, indicating synergy between these Hox factors (Barrow and Capecchi, 1999).

Control of *Hox1* and *Hox2* genes is exquisitely sensitive to RA levels. RA treatment of mouse embryos (before E8.0) causes rostral expansion and prolonged expression of Hoxb1 throughout the preotic hindbrain region and corresponding NCCs and a poorly defined Hoxb1/Krox20 boundary (Morriss-Kay et al., 1991). In the mouse, Hoxb1, Hoxb2, and Krox20 expression shows a rapid response to RA exposure, resulting in transformation of r2 and r3 to r4 and r5 identities (including a shift to more posterior Hox markers) and transformation of the trigeminal motor nerve to a facial nerve identity (Marshall et al., 1992). In chicken embryos, RA treatment of r4 at the time of hindbrain segmentation (just before NCC migration) results in loss of Hoxb1 and ectopic Krox20 in r4, altering some neurogenic NCC migration pathways but not affecting mesenchymal NCCs (Gale et al., 1996). In the developing r6 and r7 regions of compound *RAR*α-null; *RAR*β-null mutants, many patterning changes were detected: ectopic Hoxb1 expression, upregulated Hoxb3, loss of Hoxd4, and expanded MafB and Krox20 domains (Dupe et al., 1999). Similarly, *Raldh2*-null mice, instead of forming a defined r4, have Hoxb1-expressing cells spreading throughout the caudal hindbrain, and *Hox3* and *Hox4* genes, markers of r5–r8, are downregulated (Niederreither et al., 2000). In addition to RA, fluconazole and other triazole derivatives can disrupt morphogenesis of the pharyngeal arches and the cranial nerves in mouse, rat, and frog embryos, altering Hoxb1, Hoxa2, and Krox20 expression and affecting hindbrain segmentation and cranial NCC migration (Menegola et al., 2004; Menegola et al., 2005; Papis et al., 2007). RA treatment produced a phenotype similar to that of ectopic Hoxb2 expression, including ectopic Krox20 expression in the eye and fusion of cartilages from pharyngeal arches 1 and 2 (Yan et al., 1998).

7.1.2 Hoxa2 and Hoxb2

Most vertebrates have only two *Hox2* paralogous group genes, *Hoxa2* and *Hoxb2*, which share some redundant function. Mouse Hoxa2 is expressed in cranial NCCs that migrate into the second pharyngeal arch and is essential for proper patterning of NC-derived structures in this region (*Hoxa2*-null embryos have defects in second arch patterning) (Hunter and Prince, 2002; Maconochie et al., 1999). Similarly, in *Xenopus*, *Hoxa2* knockdown induces homeotic changes of the second arch cartilage (caused by ectopic upregulation of Bap, needed for jaw formation). In both mice and frogs, Hoxb2 is normally downregulated in the second pharyngeal arch and may not play a major role in second arch development (Baltzinger et al., 2005). In zebrafish, Hoxb2 is expressed at high levels in r3 and decreases in a gradient until it is undetectable in r6, but then appears again at low levels in r7and r8 (Yan et al., 1998). In zebrafish, Hoxa2 and Hoxb2 are involved in hyoid patterning, but knockdown of either gene alone has no effect (Baltzinger et al., 2005; Hunter and Prince, 2002). Knockdown of both zebrafish *Hox2* genes results in major defects in second arch-derived cartilages (Hunter and Prince, 2002). Overexpression of Hoxb1 induces ectopic expression of zebrafish *Hox2* genes in the first arch, resulting in homeotic transformations: second arch identities are formed where first arch structures normally occur (Hunter and Prince, 2002). Overexpression of Hoxb2 in early embryos resulted in abnormal morphogenesis of the midbrain and rostral hindbrain, abnormal patterning in r4, fusion of NC-derived cartilage elements arising from pharyngeal arches 1 and 2, and corresponding ectopic expression of Krox20 and MafB, but not Hoxb1 (similar to RA treatment defects) (Yan et al., 1998).

Hoxa2 transcription is controlled in part by Krox20. There is an r3/r5 enhancer in the *Hoxa2* gene with Krox20 binding activity. Mutation of the Krox20 binding sites abolishes r3/r5 activity but does not affect expression in NC and mesodermal components. Ectopic expression of Krox20 in r4 transactivates a *Hoxa2/lacZ* reporter transgene in this rhombomere (Nonchev et al., 1996). Like Hoxa2, the segment-restricted upregulation of Hoxb2 is controlled by Krox20, which is expressed in r3 and r5 (Nonchev et al., 1996; Wilkinson, 1993). The rostral boundary of Hoxb2 expression occurs at the prospective r2/r3 boundary, coinciding with the boundary of the r3expression domain of Krox20 (Ruberte et al., 1997). One NCC enhancer element in the *Hoxa2* locus contains AP-2 binding sites necessary for NCC-specific expression and can be transactivated in vitro by AP-2 proteins (Maconochie et al., 1999). Zebrafish *AP-2* ^{lockjaw} mutants have normal gene expression in the mandibular arch, but severe reductions of Hoxa2 and Dlx2 expression in the hyoid arch, suggesting a homeotic transformation to a mandibular fate (Knight et al., 2004; Knight et al., 2003). Similarly, in *AP-2*$^{mont\ blanc}$ mutants, the craniofacial primordia in pharyngeal arches two to seven fail to express their normal repertoire of genes (*Sox9a*, *Wnt5a*, *Dlx2*, *Hoxa2*, and *Hoxb2*) (Barrallo-Gimeno et al., 2004). Binding sites within the *Hoxb2* promoter for Hox cofactors such as Pbx and Meis factors, are necessary for Hoxb2 expression in r3, r4, and r5 (Scemama et al., 2002). Knockdown of Pknox1, another Hox cofactor, severely affects hindbrain segmentation and patterning, and results in loss or defective expression of Hoxb1, Hoxa2 and Hoxb2) (Deflorian et al., 2004). The nuclear matrix protein Satb2 represses Hoxa2 expression and acts with other regulatory proteins to promote osteoblast differentiation, integrating patterning and differentiation of cranial NCCs (Ellies and Krumlauf, 2006).

7.1.3 Hoxa3, Hoxb3, and Hoxd3

In vertebrates, there are three *Hox3* genes: *Hoxa3*, *Hoxb3*, and *Hoxd3*. Hoxa3 is necessary for thymus development and is expressed in cells of the third and fourth pharyngeal arches, which are involved in thymus organogenesis. Hoxa3 expression in the NT and caudal pharyngeal arches quickly increases in embryos treated with RA, and subsequent changes in Crabp expression and cranial ganglia morphology are consistent with altered NCC migration (Mulder et al., 1998). Hoxb3 is expressed in the pharyngeal arches, NT, somites, cranial ganglia, NC, and GI tract (Sham et al., 1992). *Hoxb3*-null mice have minor defects of the cervical vertebrae and the ninth cranial nerve (Manley and Capecchi, 1997). *Hoxa3*-null and *Hoxd3*-null mutant mice have no overlapping defects, but mice mutant for both genes have new defects (such as loss of the atlas) suggesting these two genes strongly interact. All three members of this group interact synergistically to affect development of both neuronal and mesenchymal NC-derived structures, and somitic mesoderm-derived structures. *Hoxa3*-null; *Hoxd3*-null compound mutants are indistinguishable from *Hoxb3*-null; *Hoxd3*-null compound mutants, indicating the significance of *Hox3* gene dosage (Manley and Capecchi, 1997). *Hox3* genes, along with *Hox4* and *Hox5*, also contribute to cardiac NCC development; knockdown of *Hox3* gene expression causes regression of aortic arch arteries (Conway et al., 1997b; Kirby et al., 1997). When r5 and r6 are surgically ablated in the chicken embryo, Hoxa3-expressing cells from r7 enter the third pharyngeal arch normally. However, some r4 cells migrate abnormally into the third pharyngeal arch and upregulate Hoxa3, a transcript they do not normally express (Saldivar et al., 1997). Transplantation of r4–r6 in a reverse orientation misdirects the NC and results in a reduction of Hoxa3 expression in the third pharyngeal arch and corresponding defects in third arch-derived structures of the hyoid apparatus (Saldivar et al., 1997).

7.2 Hox4 AND Hox5 PARALOGOUS GROUPS

In mice and humans, *Hox4* and *Hox5* genes are expressed at an anterior–posterior level roughly corresponding to the sites of origin of vagal NCCs in humans and mice. Vertebrates have four *Hox4* genes and three *Hox5* genes. Hoxa4, Hoxb4, Hoxc4, Hoxd4, Hoxa5, Hoxb5, and Hoxc5 have different expression patterns that correspond to different morphological regions of the developing ENS, forming a specific enteric Hox code necessary for enteric development (Pitera et al., 1999). Like other *Hox* genes, Hoxd4 expression is responsive to RA; the *Hoxd4* locus has a 3 enhancer containing an RARE required for initiation, maintenance, and anteriorization of a *Hoxd4* transgene in the neurectoderm (Zhang et al., 2000). Inactivation of Raldh2 in zebrafish leads to delayed and significantly reduced expression of Hoxb4 in the nervous system (Begemann et al., 2001). A chimeric fusion protein of Hoxb5 with the Engrailed repressor domain in place of the normal Hoxb5 transcription activation domain competes with endogenous Hoxb5 for target binding, causing a dominant-negative effect. NCCs expressing this repressor fusion protein fail to migrate to the distal intestine, resulting in reduction or absence of ganglia corresponding to a significant reduction or absence of Ret expression in NCC and ganglia (Lui et al., 2008). Hox4 and Hox5 also contribute to cardiac NCC development. Knockdown of Hox5 expression results in an additional pharyngeal arch containing a novel, independent aortic arch artery (Kirby et al., 1997).

7.3 POSTERIOR Hox GENES (Hox6, Hox7, Hox8, AND Hox9 PARALOGOUS GROUPS)

Whereas Hoxb1–Hoxb5 are sensitive to RA but not FGF signaling, Hoxb6–Hoxb9 do not respond to RA treatment but can be expanded anteriorly in the NT to the level of the otic vesicle with FGF treatment mediated by Cdx activity. Increased Cdx activity combined with FGF treatment causes anterior expansion of Hoxb9 expression (Bel-Vialar et al., 2002). These posterior *Hox* genes still play important roles in NC patterning and development. Lineage labeling from the $\textit{Hoxb6}^{CreERT}$ knock-in mouse labels NC progenitors (Nguyen et al., 2009). Specific *Hoxa7* enhancer elements are sufficient to drive expression of a reporter gene in major areas of the PNS (Tremblay et al., 1995). Exposure to the teratogen arsenate at E9 causes delayed NT closure, downregulation of Hoxc8, and upregulation of Pax3, both regulators of Ncam1 expression, possibly resulting in compromised NCC migration (Wlodraczyk et al., 1996). Mouse trunk NCCs have chondrogenic potential and FGF2 is an inducing factor for chondrogenesis in vitro. Fgf2 altered the expression patterns of *Hox9* genes and *Id2*, a cranial NCC marker. These data support the hypothesis that environmental cues play essential roles in generating differences between developmental patterns of cranial and trunk NCCs (Ido and Ito, 2006). Posteriorly restricted expression of these genes is also critical for proper development. Transgenic mice that express Hoxa7 ubiquitously have craniofacial defects and die shortly after birth. RA treatment produces similar results, suggesting a common mechanism (Balling et al., 1989).

The role of members of the Hox10–Hox13 paralogous groups in NC development has not been examined in great detail and so will not be reviewed here, but the Hox11-related homeobox gene *Tlx1* is reviewed in the following chapter.

• • • •

CHAPTER 8
Lim Genes

8.1 Lhx1 AND Lhx7

LIM (Lin11/Isl1/Mec3) domain proteins are transcription factors containing a unique zinc finger motif often involved in protein–protein interactions with other transcription factors. One subclass of LIM homeodomain-containing proteins is encoded by *Lhx* genes. In *Xenopus* and in mice, Lhx1 (formerly called Lim1) is expressed in the developing notochord, pronephros, CNS, retina, otic vesicle, and in NC derivatives such as in the DRGs and adrenal glands (Karavanov et al., 1996). Lhx1 expression is specifically blocked by Hdac8 to control patterning of the skull. Conditional deletion of Hdac8 in cranial NCCs causes perinatal lethality due to skull instability (Haberland et al., 2009). In mammals, all NC-derived ectomesenchymal cells are equally competent to respond to Fgf8, but their fates are restricted based on proximity to the Fgf8 signal. Lhx7 marks the restricted rostral domain (Tucker et al., 1999).

8.2 Lims1, Jub, Pdlim5, Lmo4

Lims1/Pinch (particularly interesting cysteine and histidine rich protein) has five protein-binding LIM domains, is also expressed in NCCs (both in the neural folds and OFT) and myocardium, and is involved in cardiac NC and heart development in the chicken embryo. Lims1 upregulation in NCC explants suggests that Lims1 may be a regulator of NCC adhesion and migration (Liang et al., 2009; Martinsen et al., 2006). The Ajuba (Jub) family of LIM proteins acts as functional corepressors of the Snail family by interacting with the conserved SNAG repression domain of Snail. During NC EMT in *Xenopus*, Jub, which interacts with Snail on the *Cdh1* (E-cadherin) promoter to augment repression, is required for in vivo Snail/Slug function. Jub is also a component of adherens junctions and contributes to their assembly or stability (Langer et al., 2008). The Pdlim5/enigma homolog (Enh) protein is a LIM-domain containing factor that assists in regulation of Id2 activity during neural differentiation by binding the HLH domain of Id2 and retaining Id2 in the cytoplasm. Pdlim5 is upregulated during neural differentiation, and Pdlim5 ectopic expression in neuroblastoma cells causes translocation of Id2 to the cytoplasm. Knockdown of Pdlim5 prevents cytoplasmic relocation of Id2 in neuroblastoma cells differentiated with RA. NC-derived differentiated ganglioneuroblastomas coexpress Id2 and Pdlim5 in the cytoplasm (Lasorella and Iavarone, 2006). Nuclear LIM domain-only (Lmo) proteins consist of two zinc finger motifs that mediate interactions between various transcription factors. Lmo4 is expressed in Schwann cell progenitors after neurite contact and is essential for proliferation of neuroepithelial cells in the anterior NT. *Lmo4-null* embryos have failure of anterior NT closure and die in utero (Lee et al., 2005a).

• • • •

CHAPTER 9
Pax Genes

Vertebrate *Pax* genes are related to the *Drosophila* paired-rule gene, *paired*, which encodes a protein with two DNA binding domains, a paired domain and a paired-like homeodomain.

9.1 Pax3

Pax3 is the most well-characterized Pax protein in the NC. It is also one of the most well-characterized transcription factors of the NC and has been extensively studied in NC induction and cardiac NC and melanocyte lineages. Progress in understanding the function of Pax3 has been aided by a number of mutations in mice and humans.

9.1.1 Pax3 in Early Neural Crest Development and Migration

In the mouse, Pax3 is first detected at E8.5 in the dorsal neuroepithelium and adjacent to the segmented dermomyotome (Figure 9.1). Pax3 is expressed during early migration of NCCs and in somitic cells along the NCC migratory path (Serbedzija and McMahon, 1997). Between E10 and

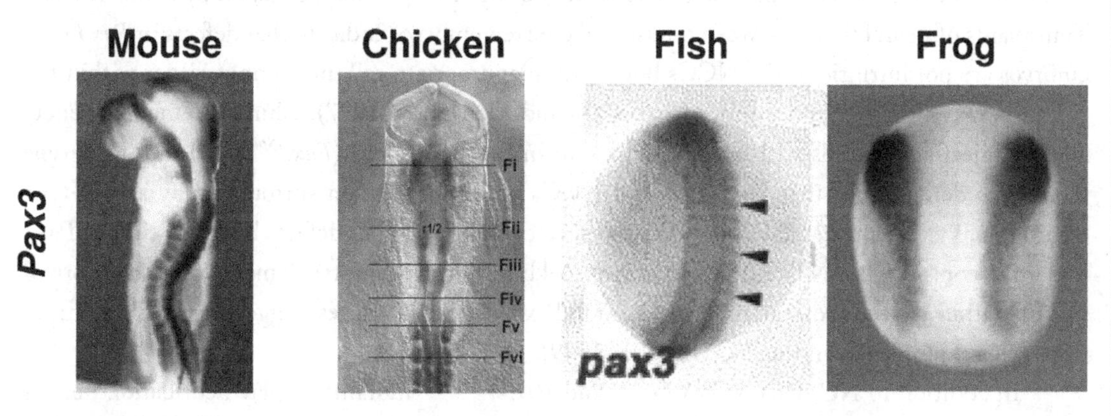

FIGURE 9.1: Whole-mount in situ hybridizations show *Pax3* in mouse, chicken, fish (*pax3a*), and frog embryos. Mouse: Solloway and Robertson, 1999, 12-somite embryo, dorsal view; chicken: Bothe and Dietrich, 2006, HH10 stage embryo, dorsal view; zebrafish: Minchin and Hughes, 2008, 15-somite embryo, dorsal view; *Xenopus*: Sato et al., 2005, early neurula embryo, dorsal view.

E12, Pax3 is detected in NCCs of the developing PNS, the NC-derived craniofacial mesenchyme, and the migratory cardiac NCCs (Goulding et al., 1991). Pax3 is generally extinguished in migratory NCCs as they differentiate, but maintained in the melanocyte lineage and also sustained in cultured NCSCs [such as multipotent SKPs (Zhao et al., 2009)] in later stages of development in various lineages. The timing of expression in early NC and subsequent downregulation as NC-derived cells differentiate also implicate Pax3 as an important factor for maintaining progenitor populations. In fact, persistent Pax3 misexpression in cranial NCCs resulted in cleft palate and other craniofacial defects, ocular defects, and perinatal lethality (Wu et al., 2008). One phenotype associated with persistent Pax3 expression is that BMP-induced osteogenesis is blocked via upregulation of the Pax3 target Sostdc1, a soluble BMP inhibitor. This is one example by which Pax3 plays a role in maintaining an undifferentiated state by blocking responsiveness to differentiation signals (Wu et al., 2008).

One of the reasons Pax3 function is well-characterized in the NC is the availability of a series of mutations in the murine *Pax3* gene. These alleles represent a range of null, severely hypomorphic, and mildly hypomorphic forms of Pax3. Embryos homozygous for the *Pax3*-null mutation $Pax3^{Splotch}$ have NC defects (Henderson et al., 1997; Tremblay et al., 1995) including reduction or loss of spinal and sympathetic ganglia, cranial ganglia defects (Franz and Kothary, 1993; Tremblay et al., 1995), pigmentation defects, and cardiac outflow tract (OFT) septation failure (Henderson et al., 1997). $Pax3^{Splotch}$ NCCs fail to fully colonize target tissues because of a severe reduction in the number of NCCs that emigrate from the neural tube (NT) at the vagal and rostral trunk levels and a complete loss of cells at the caudal thoracic, lumbar, and sacral levels. There are conflicting data in the literature concerning the cell autonomy of the NC phenotype caused by a loss of Pax3. Transplants of neural tissue between mouse and chicken embryos indicate that defects in $Pax3^{Splotch}$ embryos are not intrinsic to the NCCs but reflect inappropriate cell interactions either within the NT or between the NT and somites (Serbedzija and McMahon, 1997). Chimeric embryos generated by injecting ES cells with a LacZ knock-in in the Pax3 locus ($Pax3^{Splotch2G}$) into wild-type blastocysts demonstrate that Pax3-deficient NCCs are rescued when surrounded by a wild-type environment (Mansouri et al., 2001). However, rescue of $Pax3^{Splotch}$ defects by expression of Pax3 under the control of an NT and NC-specific 1.6-kb minimal promoter demonstrates that normal NC migration and function can occur adjacent to Pax3-deficient somites, suggesting that Pax3 functions cell autonomously in the NC (Li et al., 1999).

In addition to NC defects, $Pax3^{Splotch}$ and $Pax3^{Splotch2H}$ mutants display neurulation defects and altered somitogenesis. One link between these three tissue types (NC, NT, and somites) is the importance of extracellular matrix (ECM) composition and expression of cell adhesion molecules. Versican, a chondroitin sulfate proteoglycan nonpermissive for NCC migration in vitro, is upregulated in the NCC migration path of $Pax3^{Splotch2H}$ embryos. Pax3 and versican normally have mutually exclusive expression; and when Pax3 is lost, the versican expression domain is expanded, correlating with an absence of migrating NCCs lateral to the NT and in the lower pharyngeal arches (Henderson et al., 1997). Ncam, another Pax3 target, depends on the attachment of polysialic acid (PSA) for its adhesive properties. *Pax3^{Splotch2H}* homozygotes have a reduction of PSA-Ncam starting around E12.5, which could contribute to decreased NC migration (Glogarova and Buckiova, 2004), and the St8sia2 (α-2,8-polysialyltransferase) gene (also known as Stx in humans) is also a Pax3 target (Mayanil et al., 2001).

A wealth of knowledge regarding NC induction has been gained from studies in frog and chicken. In *Xenopus*, Msx1 can induce multiple early NC genes including Pax3 and Zic1 (Monsoro-Burq et al., 2005), which are among the first genes expressed in response to neural plate border-inducing signals (Davidson and Keller, 1999). Where Pax3 and Zic1 expression overlap, they act together (before other early NC marker genes like Foxd3 and Slug are expressed) to specify the neuroectoderm to adopt a NC fate (Hong and Saint-Jeannet, 2007; Sato et al., 2005) and activate Slug expression in a Wnt-dependent manner (Monsoro-Burq et al., 2005). Multiple experiments demonstrate that Pax3 and Zic1 together are both necessary and sufficient to specify NC (Hong and Saint-Jeannet, 2007; Sato et al., 2005). This process requires Fgf (Monsoro-Burq et al., 2005) and Wnt signaling (Bang et al., 1999; Sato et al., 2005); a dominant-negative version of Wnt8 blocks expression of both Pax3 and Msx1, and injection of Wnt8 results in an expansion of Pax3 expression in the lateral neural plate (Bang et al., 1999). In addition, expression of eIF4AII in *Xenopus* animal cap explants causes upregulation of Pax3 and other neural plate border genes such as Snail and Slug (Morgan and Sargent, 1997). In the chicken embryo, Pax3 is expressed in the dorsal NT and plays a role in specification of NCC (Osorio et al., 2009). Pax3 is upregulated in the dorsal NT in response to Wnt signaling but also depends on BMP signaling. By the start of NC migration, Pax3 begins to be downregulated (Burstyn-Cohen et al., 2004; Taneyhill and Bronner-Fraser, 2005).

Exploring the lineage and function of Pax3-expressing cells has been aided by the generation of genetic tools in the mouse. A 1.6-kb NT/NC minimal promoter fragment directs NC expression of Cre recombinase in transgenic mice (Li et al., 2000), allowing for fate-mapping of NC derivatives in combination with the *R26R^{lacZ}* reporter mouse. *Pax3^{Cre}*, a knock-in of Cre recombinase into the *Pax3* locus, resulted in a new *Pax3*-null allele with homozygous defects indistinguishable from *Pax3^{Splotch}* homozygotes. *Pax3$^{Cre/+}$* mice were used to fate-map all derivatives of endogenous Pax3-expressing cells, confirming known NC and somitic derivatives in addition to uncovering a few previously undetected lineages (Engleka et al., 2005). A hypomorphic allele, *Pax3^{neo}*, results in an 80% reduction of Pax3 protein in homozygotes, and reduced limb and tongue musculature due to increased apoptosis within somites, and postnatal mortality due to an inability to suckle. However, the heart, diaphragm, trunk musculature, various NC-derived lineages, and NT were all unaffected by these partially reduced Pax3 levels. Elevated levels of the related Pax7 protein were present in unaffected neural tube and epaxial somatic components, suggesting redundancy in protein function and links in gene regulation (Zhou et al., 2008a).

9.1.2 Pax3 in Neural and Glial Lineages

After Pax3 expression is extinguished in migratory NCCs, it becomes expressed once more as they condense to form DRGs and a subset of the cranial ganglia (Baker et al., 2002). While Pax3 is detected in placodal-derived cranial neurons, Pax3 is not detected in NC-derived neurons in these ganglia, suggesting either that Pax3 is only initially expressed in glial cells or that Pax3 expression is turned off completely before neuronal differentiation proceeds (Baker et al., 2002). Pax3 is also expressed in NC-derived Schwann cell (SC) precursors when they migrate to the PNS (Kioussi et al., 1995). During SC development, Pax3 function is modulated by Sox10, a key regulator of glial development (Kuhlbrodt et al., 1998a). In developing sciatic nerves, there is an inverse correlation between expression of Pax3 and myelin basic protein (Mbp) both in vivo and in SC primary cultures. Injection of Pax3 expression vector causes a decrease in Mbp and Pax3 represses a 1.3-kb Mbp promoter fragment in cotransfection assays, suggesting that it represses Mbp transcription directly (Kioussi et al., 1995) as one way of preventing differentiation of SC precursors to myelinating SCs (Jessen and Mirsky, 1998; Kioussi et al., 1995).

In the absence of functional Pax3 (such as in *Pax3^{Splotch}* mutants), the NCCs migrating to the PNS undergo premature neurogenesis (evidenced by increased Brn3 positive staining in neural tube explants), perhaps due to a change in the regulation of genes such as *Hes1* and *Ngn2* (needed for differentiation and proliferation) by directly binding to their promoters. In this role, Pax3 may couple migration with NCSC maintenance and neurogenesis (Nakazaki et al., 2008). In vitro, a different phenomenon is seen. Pax3 is initially expressed in all NCCs from culture but is subsequently only retained in neurons. *Pax3^{Splotch}* homozygous NC cultures had an 80% reduction in the capacity to generate sensory-like neurons. Downregulation of Pax3 in DRG cultures inhibited 80–90% of newly generated sensory neurons but had no effect on survival of sensory neurons or precursors (Koblar et al., 1999).

9.1.3 Pax3 in the Cardiac Neural Crest

Pax3 expression marks migrating cardiac NCCs in the mouse embryo, but Pax3 expression is extinguished before NC population of the heart. Cardiac NCCs that populate the heart are indeed generated from Pax3-expressing precursors (Engleka et al., 2005; Epstein et al., 2000; Li et al., 2000). Mutation of Pax3, as occurs in *Pax3^{Splotch2H}* homozygotes, results in development of heart defects including persistent truncus arteriosus (PTA), signs of cardiac failure (Conway et al., 1997a; Conway et al., 1997b; Conway et al., 1997c), and defects of the aortic arches, in addition to thymus, thyroid and parathyroid defects, ultimately leading to embryonic lethality. In the developing thymus and parathyroid, Pax3-expressing NCCs contribute to patterning of the endoderm of the third pouch into appropriately fated thymus vs. parathyroid domains (Conway et al., 1997b; Griffith et al., 2009). Between 60% and 85% of *Pax3^{Splotch2H}* homozygotes die at E13.5–E14.5 (Conway et al., 1997a; Conway et al., 1997c). Closely related *Pax3^{Splotch}* homozygotes also die by E13.5 with approximately the same spectrum of cardiac defects (Epstein et al., 2000). Pax3-expressing NCCs emigrate from the NT of both *Pax3^{Splotch}* and *Pax3^{Splotch2H}* homozygotes, but there is a reduction of total NCCs (as indicated by multiple markers, including AP-2) that pass through the pharyngeal arches and into the cardiac OFT (Conway et al., 2000; Conway et al., 1997b; Epstein et al., 2000). NC abnormalities in *Pax3^{Splotch2H}* homozygotes result in defective excitation-contraction coupling, as indicated by a significant reduction in Ca2+ transients. There is a correlation between the presence of OFT septation defects (~60–85%) and defects in other NC lineages, such as reduction or absence of DRGs (Conway et al., 1997c). Embryos without a PTA survive until birth but then they die of neural tube defects (NTDs). These embryos do not have excitation–contraction coupling defects (Conway et al., 1997a) and have less severe defects in other NC lineages like the DRG, indicating that excitation coupling and earlier lethality were secondary to an underlying NC-related heart defect (Conway et al., 1997a; Conway et al., 1997c). The reduction in NCC number in *Pax3^{Splotch2H}* homozygotes is not due to impaired migration or effects on proliferation and apoptosis in migrating NCCs but rather due to a failure of the NCCs to proliferate while still in the neural folds (before the onset of migration), which corresponds with a decrease in Wnt1 levels (Conway et al., 2000). In *Pax3^{Splotch}* homozygotes, defects are again not due to disruption of cardiac NC migration per se but rather to a reduction in the number of initial progenitors and an alteration of more finely tuned migratory behavior in mutant cardiac NCCs (Epstein et al., 2000). The reduction in the initial pool of cardiac NCC progenitors is due, at least in part, to Pax3 regulation of p53-dependent apoptosis. Pax3 inhibits p53-dependent apoptosis in the dorsal NT to regulate neural tube closure (Pani et al., 2002) and is also important for normal development of the cardiac NC. In *Pax3^{Splotch}* nulls, there were fewer cardiac NCCs due to an increase in cell death as migration occurred. Loss or suppression of p53 prevents defective CNC migration and apoptosis in *Pax3^{Splotch}* homozygotes and restored proper OFT septation (Morgan et al., 2008a).

Expression of Pax3 in the NT and NC is also affected by environmental conditions. Oxidative stress (such as that induced by hyperglycemia in diabetic mothers) can somewhat inhibit the expression of Pax3 in the neuroepithelium (Kumar et al., 2007; Morgan et al., 2008b), and this correlates with an increased risk of congenital heart and neural tube defects observed with gestational diabetes (Kumar et al., 2007; Morgan et al., 2008b). In these embryos, few cardiac NCCs migrated to the heart and p53-mediated apoptosis of progenitor cells increases along the normal migration path (Chappell et al., 2009; Morgan et al., 2008b). Along with reduced proliferation (Kumar et al., 2007), this resulted in a significant increase in OFT and other defects (Kumar et al., 2007; Morgan et al., 2008b). These defects could be ameliorated by addition of antioxidants before Pax3 expression (Morgan et al., 2008b). Folic acid deficiencies, an independent environmental stress, are associated with increased risk for congenital heart defects, and impaired folic acid transport results in extensive apoptosis in the OFT and interventricular septum correlating with a significant reduction in Pax3 expression in the presumptive migrating cardiac NC (Tang et al., 2004). Similar to the reduction of Pax3 expression due to hyperglycemic oxidative stress, cardiac NC defects likely arise due to a reduction in the size of an initial progenitor pool. Similarly, administration of nitrofen in rats causes a reduction in Pax3 expression, changes in NC signaling, and heart defects, among other effects, and both Pax3 expression and the heart defects can be mostly recovered by administration of vitamin A (Gonzalez-Reyes et al., 2005; Gonzalez-Reyes et al., 2006). *Pax3* is also one of the genes upregulated in the NT in response to arsenate treatment associated with an increase in NT closure and NCC migration defects (Wlodraczyk et al., 1996).

9.1.4 Pax3 in the Melanocyte Lineage

A subset of the NC-derived cells that emigrate from the NT travels through the dorsolateral pathway and becomes committed to the melanoblast lineage; these cells maintain Pax3 expression. Pax3 expression in the melanocyte lineage is conserved in mammals and birds (Lacosta et al., 2005). Pax3 is one of the factors in a network (along with Sox10 and Mitf) crucial for melanocyte stem cell survival and regulation of differentiation and melanin synthesis (Boissy and Nordlund, 1997; Sommer, 2005). Pax3 directly binds to and activates the *Mitf* and *Trp1* promoters (Corry and Underhill, 2005; Lang and Epstein, 2003). When Sox10 is independently bound to the *Mitf* promoter, Mitf transcription is synergistically activated. The downstream genes *Tyr*, *Tyrp1*, and *Dct*, essential for melanin synthesis, have binding sites for Pax3, Mitf, and Sox10 (Corry and Underhill, 2005; Murisier and Beermann, 2006), indicating that a complex interaction network maintains the balance between melanocyte stem cells and differentiated melanocytes. Indeed, although Pax3 activates Mitf expression, it also competes with Mitf for occupancy of the *Dct* promoter and represses Dct expression. Pax3 thus maintains the undifferentiated state of these cells until repression by Pax3 is relieved through activation of the Wnt signaling pathway (Lang et al., 2005). Analysis of *Pax3^{Splotch}* embryos also reveals a significant reduction in the number of melanoblasts indicating a role for Pax3 in expanding the pool of restricted progenitor cells by regulating Mitf, which is necessary for melanoblast survival (Hornyak et al., 2001). Interestingly, Pax3 is one of the genes greatly reduced in white or graying hairs, further reinforcing its role in melanocyte stem cell populations (Choi et al., 2008). This role for Pax3 in pigment cell development is conserved in zebrafish. Pax3 MO knockdown results in defective fate specification of xanthophores, but the other two pigment lineages, melanophores and iridophores, are specified and differentiate normally. Loss of xanthophores is likely due to a Pax3-driven fate switch within a pigment cell precursor population (Minchin and Hughes, 2008). Foxd3 can act to inhibit Pax3 activation of Mitf by directly binding the Pax3 protein and sequestering it from binding to the Mitf promoter (Thomas and Erickson, 2009).

Another mechanism for regulation by Pax3 is differential expression of isoforms. In mice and humans, seven alternatively spliced isoforms (a, b, c, d, e, g, and h) of Pax3 have been described, with Pax3c being the most well-characterized isoform. In mouse melanocytes in vitro, the effect of Pax3 on proliferation, migration, survival, and transformation varies depending on the isoform, with some isoforms having opposite effects (Wang et al., 2006). Microarray experiments suggest that these varying effects are due to differential regulation of distinct yet overlapping sets of genes involved in cell differentiation, proliferation, migration, adhesion, apoptosis, and angiogenesis by the various isoforms (Wang et al., 2007).

9.1.5 Pax3 in Waardenburg Syndrome, Tumors, and Other Neurocristopathies

Waardenburg syndrome (WS) patients have an autosomal dominant combination of auditory and pigment defects. This is due to a loss of melanocytes from the skin, hair, eyes, and cochlea of the inner ear, reminiscent of an NC deficiency (Read and Newton, 1997) and remarkably similar in phenotype to the *Pax3*^{Splotch} heterozygous mutant mouse. Multiple studies identified Pax3 mutations in families with either type 1 or type 3 WS (Karaman and Aliagaoglu, 2006; Ptok and Morlot, 2006; Read and Newton, 1997; Tassabehji et al., 1993; Tassabehji et al., 1994; Tassabehji et al., 1995; Van Camp et al., 1995). These mutations typically alter conserved amino acids in either of the DNA binding domains or cause a loss-of-function via deletion, nonsense, splice-site, or frameshift mutations (Tassabehji et al., 1993; Tassabehji et al., 1994; Tassabehji et al., 1995), and there is usually a close correspondence in humans and mice between the type of Pax3 molecular lesion, gene dosage, and strength of observed phenotype (Corry et al., 2008; Tassabehji et al., 1994). A subset of type 2 WS cases has been linked to mutations in Mitf (Lalwani et al., 1998; Potterf et al., 2000) and Pax3 enhances activation of Mitf by Sox10 (Potterf et al., 2000). Misregulation of Mitf due to mutation of Pax3 in type 1 and type 2 WS may also cause at least part of the WS phenotype (Potterf et al., 2000). Some patients with WS also have characteristics of Hirschsprung disease, and Pax3 is required for normal enteric ganglia formation. Pax3 can bind to and activate expression of Ret in coordination with Sox10, and mutations in both have been associated with Waardenburg–Hirschsprung syndrome (Lang et al., 2000; Lang and Epstein, 2003). This is reflected in the role for Pax3 in both pigment cells (xanthophores) and ENS in zebrafish (Minchin and Hughes, 2008). Interestingly, Pax3 mutations were typically not found in families with other neurocristopathies (Tassabehji et al., 1995).

Because of its critical role during development, especially in progenitor cell populations, it is not surprising that Pax3 also plays a role in tumor formation. Many human neuroectodermal tumors express Pax3 (Gershon et al., 2005; Parker et al., 2004; Scholl et al., 2001; Schulte et al., 1997). Pax3 was detected in combination with Pax7 in poorly differentiated tumors and tumors with malignant potential (Gershon et al., 2005). Pax3 is specifically detected in most NC-derived Ewing's family tumors (Schulte et al., 1997) and in most peripheral neuroectodermal tumors (Schulte et al., 1997), small cell lung cancer (Parker et al., 2004), and in most primary cultured melanomas (He et al., 2005; Parker et al., 2004; Scholl et al., 2001). Pax3 expression was specific to tumor cells and not detected in surrounding normal tissue or in benign lesions, and downregulation of Pax3 expression in these tumors resulted in apoptosis of primary melanoma cells (He et al., 2005; Scholl et al., 2001), suggesting a role for Pax3 in melanoma cell survival.

9.1.6 Pax3 Regulation and Interaction with Other Genes

In addition to genetic interactions that have been discussed in previous sections, a few direct upstream regulators of Pax3 have been identified. Tead2 and its coactivator Yap1 are coexpressed with Pax3 in the dorsal NT. Tead2 binds an NC-specific enhancer in the *Pax3* locus and activates Pax3 expression, and mutation of the Tead2 binding site prevents neural expression of Pax3. Further supporting the role of Tead2 as an activator of Pax3, a Tead2-Engrailed fusion protein represses retinoic acid-induced Pax3 expression in P19 embryonal carcinoma cells and in vivo (Milewski et al., 2004). A minimal promoter element for NT and NC expression in the Pax3 upstream regulatory sequence contains putative interaction sites that have been tested in P19 cells induced to express Pax3 by retinoic acid (RA) treatment. Two sites interact with the neural-specific genes Brn1 and Brn2, whereas other sites interact with Hox/Pbx heterodimers and Meis monomers, and all of these sites are important for normal Pax3 expression. Ectopic expression of both Brn2 and HoxA1 together induces Pax3 expression (Pruitt et al., 2004). *Pbx1*-null embryos lose a transient burst of Pax3 expression in premigratory cardiac NCCs that ultimately specifies cardiac NCC function for OFT development but does not regulate NCC migration to the heart. Pbx1 directly activates Pax3, leading to repression of its target gene Msx2 in NCCs. Compound *Msx2*-null; *Pbx1*-null embryos display significant rescue of OFT septation, demonstrating that disruption of the Pbx1–Pax3–Msx2 regulatory pathway partially underlies the OFT defects in *Pbx1*-null embryos (Chang et al., 2008). In the neural tube, Nf1 modifies *Pax3*^{Splotch} mutant defects through a genetic interaction, and both of these factors play important roles in development of NC-derived structures (Lakkis et al., 1999).

Genetic interactions in *Pax3*^{Splotch} embryos and interactions assayed in vitro have identified some downstream effectors of Pax3. The promoter activity of *Lbx1*, a gene expressed in cardiac NC during tubular heart formation, is upregulated in *Pax3*^{Splotch} mutants. *Lbx1* nulls have defects in heart looping and myocardial hyperplasia, and Lbx1 may be required for specification of a subpopulation of cardiac NC subsequent to migration (Schafer et al., 2003). *Pax3*^{Splotch} mutant embryos have reduced Lbx2 expression, as indicated by an *Lbx2*^{lacZ} knock-in allele, in DRGs and cranial nerve ganglia, but not in the urogenital system, where Lbx2 is also expressed, suggesting that Pax3 is required for Lbx2 expression in NC-derived tissues (Wei et al., 2007). Foxd3 is genetically downstream of Pax3 and is not expressed in regions of Pax3 mutant embryos lacking NC (Dottori et al., 2001). In addition, direct targets of Pax3 have been identified in several NC lineages. Wnt1 is expressed in premigratory NC, and Pax3 specifically binds the *Wnt1* promoter to regulate expression in a dose-dependent manner in the developing embryo (Fenby et al., 2008). Pax3 regulates TGFβ2, a modifier of NC migration and differentiation, by directly binding its promoter (Nakazaki et al., 2009), and loss of one allele of Pax3, such as in *Pax3^{Splotch}* heterozygotes, can reverse the TGFβ2-null NTD phenotype (Nakazaki et al., 2009). Aberrant upregulation of Msx2, a regulator of BMP signaling, in *Pax3^{Splotch}* mutant mice may lead in part to deficient migration of cardiac NC and OFT septation failure, and Pax3 represses Msx2 expression directly through a conserved Pax3 binding site in the Msx2 promoter (Kwang et al., 2002). A direct, physical protein–protein interaction has been shown between Pax3 and Hira, a transcriptional corepressor expressed in the NC that also physically interacts with core histones. This suggests that at certain targets, Pax3 may recruit Hira and exert a repressive effect on transcription via chromatin modifications. *Hira* mutants display OFT defects and Hira maps to the DiGeorge/velocardiofacial syndrome critical region 22q11. *Pax3^{Splotch}* homozygotes have many of the hallmark defects of DiGeorge syndrome (Magnaghi et al., 1998).

9.2 Pax6

The *Pax6* gene is disrupted in both mouse and rat small eye (Sey) mutants. Homozygous null *Pax6^{Sey}* rodents have eye defects similar to a group of human developmental disorders such as Peters' anomaly, Axenfeld–Rieger anomaly, and aniridia (Cvekl and Tamm, 2004), as well as craniofacial defects of the ocular and frontonasal regions (Kanakubo et al., 2006; Matsuo et al., 1993; Osumi-Yamashita et al., 1997) due to defective NCC migration (Matsuo et al., 1993; Nagase et al., 2001; Osumi-Yamashita et al., 1997). In the frontonasal region, Pax6 is strongly expressed primarily in the frontonasal ectoderm and is required non-cell-autonomously to guide NC migration (Osumi-Yamashita et al., 1997). In the eye, Pax6 appears to be a key factor in the interaction between NC-derived cells and placodal-derived cells (Cvekl and Tamm, 2004; Song et al., 2007). Pax6 is strongly expressed in the non-NC-derived epithelial cells of the anterior segment where it is required for lens induction and corneal and retinal development (Kanakubo et al., 2006), and it is also required non-cell-autonomously for migration of the NCCs that contribute extensively to the ocular mesenchyme (Kanakubo et al., 2006). The main result of Pax6 loss in the eye is failure of NCC migration through the Pax6-null surrounding epithelial cells, but two studies have shown that even though Pax6 is most predominantly expressed in the epithelial cells critical for anterior eye development (lens, cornea, retina), it is also expressed at low levels in the NC-derived ocular mesenchyme (corneal stroma, ciliary body, iridocorneal angle, and trabecular meshwork progenitors)

(Baulmann et al., 2002; Collinson et al., 2003). Weak expression of Pax6 in NC-derived cells of the developing eye is seen from P1 to P5 and becomes weaker at later stages, with no NC Pax6 detected by P14 (Baulmann et al., 2002). In the TM progenitor cells, Pax6 is downregulated upon differentiation to trabecular meshwork cells, consistent with a role for Pax6 in maintaining cells in a progenitor state.

Pax6 is also expressed in NC-derived mesenchymal cells of the submandibular gland (SMG), otherwise known as the salivary gland. In *Pax6*^{Sey} mutants, the NC-derived SMG mesenchyme exhibits defective migration (Jaskoll et al., 2002). Surprisingly, Pax6 is not detected in the epithelium, suggesting that here Pax6 has a predominantly cell-autonomous function in contrast to anterior eye development. After SMG branching morphogenesis occurs, Pax6 mesenchymal expression can no longer be detected (Jaskoll et al., 2002). Pax6 is involved in FGF-mediated branching morphogenesis of other tissues, so it may play a similar role in the SMG (Makarenkova et al., 2000).

The role of Pax6 in the eye is also likely conserved among nonmammalian vertebrates. In the chicken embryo, a population of periocular mesenchyme cells expresses Pax6 along with Lmx1b and Pitx2. This population seems to be under the control of Six3, also expressed in a subset of periocular mesenchymal cells and in differentiating anterior segment tissues. Misexpression of Six3 causes a reduction of the Pax6/Lmx1b/Pitx2-expressing mesenchymal population and disrupts the integrity of the corneal endothelium and the expression of ECM components critical for corneal transparency, leading to anterior eye defects (Hsieh et al., 2002). In *Xenopus*, administration of VPA causes a decrease in Pax6 expression in the eye and perturbation of NC migration and abnormal retinal development (Pennati et al., 2001). In zebrafish, morpholino knockdown of Olfm2 (olfactomedin-2 or OM2) results in alteration of Pax6 expression pattern in the eye and coincident disruption of ocular development (Lee et al., 2008a). Olfm2 knockdown also affects development of caudal pharyngeal arches resulting in defects in cartilaginous structures, indicating cranial NC-related defects (Lee et al., 2008a).

9.3 Pax7

In the mouse, Pax7 is expressed in the cranial NC and its derivatives during embryogenesis in addition to skeletal muscle and the CNS (Lang et al., 2003; Mansouri et al., 1996). Maintenance of Pax7 expression in the adult has also been described in NC-derived Schwann cells in the inner and outer capsules of neuromuscular spindles of hindlimb skeletal muscle (Rodger et al., 1999). Pax7 homozygous-null mice die of NC-related craniofacial defects shortly after weaning. Because Pax7 and Pax3 are highly homologous and have overlapping expression patterns, there is functional redundancy between the two genes (Mansouri et al., 1996). Complex gene regulatory interactions between Pax3 and Pax7 may also allow one of these proteins to compensate for the loss of the other; in *Pax3*^{neo} hypomorphic mice, Pax3 protein is reduced by 80%, but where Pax3 and Pax7 expression overlap, Pax7 is upregulated and no NC defects are observed (Zhou et al., 2008a).

Pax7 in the chicken embryo is expressed in NCCs, muscle precursor cells, and the developing CNS (Kawakami et al., 1997). Pax7 is one of the earliest indicators of NC precursors, along with Pax3, Snail2, Foxd3, and Sox9, with expression beginning in the dorsal neural tube immediately after cavitation (Lacosta et al., 2005; Osorio et al., 2009) and indicates a region of early NC induction. Pax7 is necessary for NC formation and coincident induction of the markers Slug, Sox9, Sox10, and HNK-1 (Basch et al., 2006). Pax7 expression is maintained in migrating trunk NC and is then sequentially downregulated first in neuronal, then glial, then melanocyte precursors (Lacosta et al., 2005). In avian species, both Pax3 and Pax7 are expressed in melanocytes, but in the mouse, only Pax3 is expressed (Lacosta et al., 2005).

The expression pattern of zebrafish Pax7 within the NC also differs slightly from that of mouse and chicken (Seo et al., 1998). There are three additional zebrafish Pax7 truncated isoforms (Seo et al., 1998). During early zebrafish embryogenesis, both cranial and trunk NC express Pax7. As in other species, Pax7 is broadly expressed in cranial NCCs. Pax7 is also expressed in the trunk NC and includes both premigratory and migratory NCCs. Pax7 is expressed in pigment precursor cells (melanophore, xanthophore, and iridophore precursors). In melanophore precursors, Pax7 expression overlaps with early melanin pigment, one difference between zebrafish and the chicken embryo (Lacosta et al., 2007). Pigment stem cells present during the larva to adult transition also express Pax7 (Lacosta et al., 2007). Pax7 is one marker of the xanthophore lineage in zebrafish, and a reduction in Pax7 is seen after morpholino knockdown of Pax3, which causes defects in xanthophore fate specification, whereas melanophores and iridophores are specified and differentiate properly (Minchin and Hughes, 2008).

Although human Pax7 is normally restricted to mesoderm and not detected in the NC (Gershon et al., 2005.), Pax7 expression was detected along with Pax3 in human neuroectodermal tumors (Gershon et al., 2005). Pax7 is disrupted by a translocation specific for alveolar rhabdomyosarcoma, similar to *Pax3/Foxo1(Fkhr)* translocations (Floris et al., 2007). Not much is known about regulation of Pax7 expression in any vertebrate species. Distinct upstream and intronic enhancer elements in the Pax7 locus have been identified that can direct expression of Pax7 in the cranial NC, facial mesenchyme, mesencephalon, and pontine reticular nucleus (Lang et al., 2003). Likewise, only minimal progress has been made in identifying functional NC targets of Pax7. One study identified 34 candidate target genes by an unbiased ChIP assay and verified Pax7 occupation and activation of sites within *Gbx1*, *Eya4*, *CntfR*, *Kcnk2*, and *Camk1d* (Hong et al., 2008a).

9.4 Pax9

Most of the investigation into the role of Pax9 in the NC has been carried out in the mouse. Mouse Pax9 is expressed in the cranial NC, the midbrain, somites, limb mesenchyme, and in foregut endoderm derivatives (Kist et al., 2007; Peters et al., 1998). Mesenchymal cells expressing Pax9 in the nose, palate, and teeth are derived from NC (Kist et al., 2007). In the mandibular arch mesenchyme, Pax9 marks the prospective sites of tooth development before morphological indicators appear. This expression is maintained in the mesenchyme of the developing teeth, and Pax9 is necessary for tooth development to proceed beyond the bud stage (Amano et al., 1999; Peters et al., 1998). Pax9 homozygous-null mice die at birth, but NC-specific deletion of a floxed allele of Pax9 using Wnt1-Cre results in mice with cleft secondary palate and lack of tooth development (Kist et al., 2007). Pax9 in the arch mesenchyme may be regulated (directly or indirectly) by the transcription factors Hand1 and Hand2. *Hand1; Hand2* compound-null mutant embryos have misregulated levels of mesenchymal Pax9 expression at E12.5 and display later mandible and tooth defects (Barbosa et al., 2007). Unlike the developing tooth, where mesenchyme originates from both NC and non-NC cells, the mesenchyme of the SMG is exclusively NC-derived (Jaskoll et al., 2002). During SMG branching morphogenesis, before budding, Pax9 is expressed in both the mandibular oral epithelium and the adjacent NC-derived mesenchyme. As the SMG undergoes epithelial thickening, Pax9 is predominantly epithelial, but some expression of Pax9 remains in the mesenchyme. A functional role for Pax9 in SMG branching morphogenesis has not been defined but likely shares some characteristics with tooth development (Jaskoll et al., 2002).

• • • •

CHAPTER 10
POU Domain Genes

The POU (Pit1/Oct2/unc-86) domain-containing transcription factor family members typically bind an octamer DNA sequence and have cell type-specific effects on differentiation. Some members of this family were originally known as Oct or Brn proteins, and those designations, when more commonly used, will be used throughout the following sections.

10.1 Pou2f1/Oct1
In the *Xenopus* embryo, Oct1 expression is detected in ectodermal and mesodermal cell lineages, and becomes progressively restricted as development proceeds, but is still detected in neuronal cells and NCCs. Oct1 may play a role in specification and differentiation of neuronal cells and NCCs (Veenstra et al., 1995). In *Xenopus* animal cap assays, high levels of injected *Oct1* mRNA downregulate Slug and upregulate Ncam, typical of neural induction, but lower amounts activate eIF4AII and Slug and suppress expression of epidermal keratin, typical of NC induction (Morgan and Sargent, 1997).

10.2 Pou3f1/Oct6
Oct6 is one of the important transcription factors, together with Sox10 and Krox20, regulating Schwann cell development (Bhatheja and Field, 2006; Jessen and Mirsky, 2002). Oct6 functions during several developmental transitions: from NCCs to Schwann cell precursors, from Schwann cell precursors to immature Schwann cells, and finally from immature Schwann cells to either myelinating or nonmyelinating Schwann cells (Jessen and Mirsky, 1998). Oct6 is coexpressed with Sox10 and these two factors function synergistically through their N-termini by binding adjacent sites in target promoters (Kuhlbrodt et al., 1998a). In addition, Oct6 transactivates the *nicotinic acetylcholine receptor (nAChR)-α3* and *nAChR-β4* promoters independently of Sox10 (Liu et al., 1999). Neither Oct6 nor Krox20 is required for initial activation of myelin gene expression, but Oct6, acting together with Krox20, is necessary for the transition from the promyelinating to myelinating stage during Schwann cell development. Oct6 is expressed mainly in promyelinating Schwann cells and then downregulated before myelination. In embryos lacking Oct6, myelination is delayed, suggesting Oct6 is necessary for the timing of this transition (Kamholz et al., 1999). Oct6 also has

a role in demyelination during regeneration after nerve trauma (Svaren and Meijer, 2008). In NC cultures, Oct6 and Erm expression appears mutually exclusive. In response to glial growth factor 2 (Ggf2), Erm-positive progenitor cells give rise to Erm-positive, Oct6-negative satellite glia, but in the presence of serum and the adenylate cyclase activator forskolin, they give rise to Erm-negative, Oct-6-positive Schwann cells (Hagedorn et al., 2000). Primary rat Schwann cells in culture and a rat Schwann cell precursor line (SpL201) express Oct6 and additional Schwann cell lineage markers. In response to forskolin treatment, these cells upregulate Oct6 and myelin gene expression. Both primary cells and SpL201cells can form myelin in the presence of axons in vitro and remyelinate CNS white matter lesions in vivo (Lobsiger et al., 2001).

10.3 Pou3f2/Brn2/Oct5

Brn2 is normally expressed at high levels in melanoblasts and melanomas but at low levels in melanocytes (Eisen, 1996; Thomson et al., 1995). Knockdown of Brn2 gene expression in melanoma cells is associated with morphological change and loss of melanocyte and NC markers, including Mitf and Tyrp, and an inability to form tumors, suggesting a role for Brn2 in determination of the melanocyte lineage and melanoma tumorigenesis (Thomson et al., 1995). In cultured melanoblasts, full Brn2 induction requires SCF, FGF2, and Edn3. Without these factors, Brn2 levels decrease to a melanocyte Brn2 level. This is accompanied by increased L-DOPA expression and melanosome maturation consistent with differentiation, suggesting a role for Brn2 as an early factor in melanoblasts that negatively regulates differentiation (Cook et al., 2003). *Brn2*, like other melanocyte-specific genes *Mitf* and *Dct*, is induced by β-catenin. Several components of the Wnt/β-catenin pathway are modified in melanoma tumors and cell lines, and involvement of Brn2 in melanoma proliferation suggests a link between Brn2 and Wnt signaling during melanoma development (Larue and Delmas, 2006).

10.4 Pou4f1/Brn3

Brn3 in both mammals and birds plays a role in sensory neuron development. In the chicken embryo, Brn3 expression is first detected in placodal and migrating NC precursors of the trigeminal ganglion, and then later in the DRGs and vestibulocochlear ganglia. After condensation of the trigeminal ganglion, Brn3-positive neurons become concentrated in the portion of the ganglion distal to the brain stem. However, most of the Brn3-expressing neurons in the trigeminal ganglia are derived from sensory placodes and not NC (Artinger et al., 1998). Sensory neurons expressing Brn3 develop from dividing precursors that differentiate within 2 days after emigration from the neural tube (Greenwood et al., 1999). NCCs undergoing premature neurogenesis, such as those in *Pax3*$^{Splotch/Splotch}$ mouse embryos, express high levels of Brn3 in neural tube explants (Nakazaki et al., 2008). In *Xenopus*, Brn3 is expressed in NC populations during early development, and later in specific cranial ganglia (Hirsch and Harris, 1997). In *Xenopus*, the secreted glycoprotein Noelin1 is expressed in the developing CNS and PNS, and Noelin1 expression in animal caps induces expression of Brn3 and Neurod1 and amplifies Noggin-induced expression of both of genes. In chicken embryos, overexpression of Noelin1 causes prolonged NC migration (Moreno and Bronner-Fraser, 2001). Brn3 expression responds to Shh signaling, which is involved in specification of trigeminal sensory neurons through induction of Neurog1 expression and promotion of NCC differentiation into trigeminal sensory neurons. A subpopulation of trigeminal NCCs expresses the Shh receptor Patched, and the number of cells expressing Brn3 increases with Shh treatment (Ota and Ito, 2003). Within the condensed DRG, Brn3 and Neurog1 initiate the sensory neurogenesis program of migrating, proliferating Foxs1-negative, Sox10-positive precursors, which entails reduction of Sox10 and acquirement of Foxs1, an early sensory neuronal marker (Montelius et al., 2007). In *Brn3*-null mice, the DRG and trigeminal ganglia exhibit many common changes in gene expression, but a subset of Brn3 target genes show increased expression only in the trigeminal ganglia, which includes cells originating from both the NC and placodal ectoderm. In wild-type trigeminal ganglia, Brn3-repressed genes are silent, but their promoters exhibit histone H3-acetylation levels similar to constitutively transcribed gene loci not observed in the DRG, suggesting that chromatin modifications play a role in cell-specific target gene regulation by Brn3 (Eng et al., 2007). Brn3 is necessary for both differentiation and survival of NCC-derived sensory neurons during development and after injury (Hudson et al., 2004; Hudson et al., 2008). p53 is expressed in a subset of Brn3-positive NC-derived cells fated for the sensory neuronal lineage. Brn3-null NCCs in culture have a significant increase in apoptosis after p53 induction, suggesting that Brn3 modulates the p53-mediated fate (a decision between differentiation and apoptosis) of these NCCs (Hudson et al., 2004). In addition, Brn3 physically interacts with different p73 isoforms in NC-derived sensory neurons. Brn3 modulates interaction of cofactor binding to the p73 C-terminus to facilitate maximal activation of the proapoptotic *Bax* promoter but not the *p21/Cip1* promoter (Hudson et al., 2008).

Generation of peripheral neurons from hESCs is achieved by coculturing hESCs with the mouse stromal cell line PA6 for 3 weeks, resulting in peripheral sensory neurons expressing Brn3 and peripherin, and peripheral sympathetic neurons expressing peripherin and Th (Pomp et al., 2005). Brn3-positive, Neurog2-positive sensory neurons are induced by BMP signals from mouse trunk NCCs in culture. Bmp4 appears to inhibit glial differentiation and allow Brn3 expression, whereas FGF2 promotes glial differentiation by negatively regulating Neurog2 (Ota and Ito, 2006).

10.5 Pou5f1/Oct4

Oct4 is a transcription factor known for its requirement in maintaining pluripotency and self-renewal of ESCs. Although Oct4 expression in the NC in situ has not been described, expression of Oct4 is observed in several distinct NCSC lines. Adult rat and human palatal-derived NC-related stem cells isolated and cultivated as neurospheres express NCSC markers and more undifferentiated, ESC-like markers such as the reprogramming factors Klf4, Sox2, Oct4, and c-Myc (Widera et al., 2009). Multipotent skin-derived precursors, which can produce both neural and mesodermal progeny in vitro, also show cooperative expression of pluripotency related genes (Oct4, Sox2, Nanog, Stat3) and NC marker genes (Zhao et al., 2009). The global molecular profile of human NCCs that develop toward the end of the first month of gestation is highly similar to that of pluripotent ESCs when compared with other stem cell populations or hNCC derivatives, and this profile includes the pluripotency markers Nanog, Oct4, and Sox2 (Thomas et al., 2008). Multipotent, self-renewing adult NCSCs from human hair follicles (Epi-SCs), distinctively different from known epithelial or melanocytic stem cells, do not express squamous or melanocytic markers but express NC and neural stem cell markers and ESC factors Nanog and Oct4 (Yu et al., 2006).

• • • •

CHAPTER 11

RAR/RXR Genes

Retinoic acid (RA) and retinoid signaling, particularly in the cranial NC, play a critical role in patterning NC derivatives. These signaling events are largely mediated by the retinoic acid receptors (RARs) and the retinoid receptors (RXRs), transcription factors that often bind their targets as heterodimers and recruit transcriptional repressor complexes. There are three *RAR* genes and three *RXR* genes conserved between humans, mice, chickens, fish, and frogs (*RARα/β/γ*, and *RXRα/β/γ*). RA binding to RAR receptors induces a conformational change in the receptor, followed by the replacement of corepressor with coactivator complexes (Cvekl and Wang, 2009). In vitro, RARs bind to response elements as heterodimers with RXRs.

11.1 RARα, RARβ, AND RARγ

In the mouse embryo, RARα is associated with NCC emigration and migration. RARα and RARβ are expressed at specific levels of the hindbrain and in the spinal cord (Ruberte et al., 1991). RA signaling acts at an early stage in primary neural development during fate determination of different regions of the neuroectoderm, and RARα expression in the posterior neuroectoderm is consistent with a role in primary neurogenesis (Sharpe and Goldstone, 2000). Null mutations of all six mouse *RAR/RXR* genes have been generated. *RARα, β* and *γ* single-null mutants have defects in only a small subset of the tissues normally expressing these receptors, but *RAR* double-null mutants have defects in most tissues, including those originating from the NC-derived mesenchyme, indicating functional redundancy or dosage dependence (Mark et al., 1998). *RARα*-null; *RARβ*-null compound mutants display defects in structures partially derived from the mesenchymal and neurogenic NC such as the thymus and post-otic cranial nerves, respectively. Lack of RARα and RARβ has no direct effect on the number and migration path of NCCs, but Mafb and Krox20 expression domains are expanded, Hoxb1 and Hoxb3 are ectopically upregulated, and Hoxd4 expression is lost (Dupe et al., 1999). Treatment of cultured mouse embryos with synthetic RARβ, acting through the endodermal cells lining the pharyngeal arches but not the NCCs or ectoderm, causes hypoplasia and fusion of pharyngeal arches 1 and 2, typical of RA-related defects. This synthetic RARβ agonist also antagonizes RARα and RARγ (Matt et al., 2003). RARα is the only receptor mediating retinoid signaling in the neurectoderm during ocular development, but all three RARs mediate the action of RA during eye morphogenesis, specifically in NC-derived periocular mesenchyme (Cvekl and Wang, 2009; Matt et al., 2008). Inactivation of RARα/β/γ receptors in the periocular mesenchyme abrogates anterior eye segment formation (Cvekl and Wang, 2009). One identified target of RARα in neuroblastoma and NC-derived cells is the developmentally regulated, retinoid-responsive thymosin β10 gene (Hall, 1992).

11.2 RXRα, RXRβ, AND RXRγ

The RARs often function in heterodimers with RXRs. Mutant embryos lacking different combinations of both RARs and RXRs indicate that RXRα:RARγ heterodimers are instrumental in patterning craniofacial skeletal elements, whereas RXRα:RARα heterodimers may be preferentially involved in generation of NC-derived arterial smooth muscle cells (Mark et al., 1998). Both RXRα: RARβ and RXRα:RARγ heterodimers function in development of the ocular mesenchyme (Mark et al., 1998).

In the trunk region, RXR transcripts are expressed by cells in the premigratory NC and in NCCs migrating through the sclerotome, indicating that NCCs express RXR transcripts before differentiation to PNS derivatives, suggesting a role for RA, mediated by RXR, in the developing PNS. Chicken RXRα is expressed at high levels during development of the NC-derived PNS, including DRG, cranial ganglia, enteric ganglia, and peripheral nerve tracts (Rowe et al., 1991). RXRα-null mouse embryos die in utero between E13.5 and E16.5 due to ventricular hypoplasia, but the NC-contributed OFT and associated vessels are normal (Sucov et al., 1994). During early chicken embryogenesis, RXRα transcripts are present in the CNS and at reduced levels in the NC and limb buds. RA upregulates RXRα transcripts at some stages but not does not expand the expression domain (Hoover and Glover, 1998). RXRγ marks migrating cranial NCCs in the chicken embryo and is gradually restricted to specific differentiating neurogenic NC derivatives such as the TG and DRGs (Rowe and Brickell, 1995, 1031). Unlike the chicken, where RXRγ is expressed from the onset of NC migration, in rat embryos, only RXRγ expression later in the DRG and TG, which is not detected until E14.5, is conserved (Georgiades et al., 1998).

• • • •

CHAPTER 12

Smad Genes

Transcription factors of the Smad family (named because of homology to *C. elegans* Sma and *Drosophila* Mad) are downstream effectors of BMP and TGFβ signaling. Phosphorylation of Smad proteins upon activation of BMP/TGFβ signaling promotes Smad localization to the nucleus, where phosho-Smads are capable of binding specific Smad sites in target gene promoters. Smad1, Smad5, and Smad8 transduce BMP signals, whereas Smad2, Smad3, and Smad9 transduce TGFβ signals. Smad4 transduces both BMP and TGFβ signals, and Smad6 and Smad7 are inhibitory Smads. Smad proteins are fairly ubiquitous, and although Smad function has not been exhaustively probed in all NC lineages, these proteins are involved in the development of most NC lineages where BMP or TGFβ signaling pathways are important.

12.1 EARLY NC

Zebrafish Smad5 mutants display a dramatic expansion of the prospective NC region supporting the hypothesis that BMP signals control the NC domain (Nguyen et al., 1998). Smad1 can bind to and activate cis-regulatory elements in the upstream sequence of the chicken *Slug* gene. Phosphorylated Smad1 is detected in the neuroectoderm and likely mediates Slug expression and subsequent EMT induced by Bmp4 (Sakai et al., 2005). TGFβ may regulate patterning and fate specification of cranial NCCs through activation of Smads and subsequent regulation of transcription factors during embryogenesis (Chai et al., 2003). In *Xenopus*, Smad3 is expressed in trunk migratory NCCs (Dick et al., 2000).

12.2 CARDIAC NC

Mice lacking the common transducer Smad4 in NCCs have PTA, hypoplastic OFT cushions (due to decreased Msx1 and Msx2 expression, leading to apoptosis), and defective OFT elongation and mispositioning (due to reduced levels of Id proteins and Mmp14) (Jia et al., 2007). Smooth muscle differentiation requires BMP-mediated Smad1/5/8 activation (Rajan et al., 2003). TGFβ activation of the SM22α promoter depends on Smad1, Smad3, or Smad4, which bind a Smad binding site and a Medea box within the SM22α promoter in the NC cell line Monc-1, a cell line that differentiates into VSMCs upon TGFβ induction. A complex of Smad3 and Smad4 together contributes to maximal promoter activity (Chen et al., 2003). Smad signaling may function in part by activating genes controlling VSMC differentiation, regulating transcription in a CArG-dependent manner (Li et al., 2007). In vitro, TGFβ repression of Fgfbp1 in mesenchymal and NCCs undergoing smooth muscle differentiation requires Smad2 and Smad3 (Briones et al., 2006).

12.3 NEURAL, GLIAL, AND SYMPATHOADRENAL NC

In contrast to smooth muscle differentiation, glial differentiation occurs at high cell densities and in response to BMP signaling (Rajan et al., 2003). In an NC-derived cell line with immature Schwann cell characteristics, Bmp2 promotes expression of the early glial marker Gfap, mediated indirectly by Smad signaling (Dore et al., 2009). During differentiation of NC-derived catecholaminergic neurons in culture, Bmp4 causes nuclear translocation of Smad1 and induces transcription of *Hand2*, required for differentiation of catecholaminergic neurons (Wu and Howard, 2001). Strongly activated cAMP signaling mediates activation of several kinases leading to cytoplasmic accumulation of phospho-Smad1, which terminates Bmp2-induced sympathoadrenal cell development (Ji and Andrisani, 2005).

12.4 OCULAR NC

Heparan sulfate deficiency in mouse NCCs causes anterior chamber dysgenesis. In mouse NCCs, disruption of Exostosin1, an enzyme necessary for heparan sulfate synthesis, results in disruption of Tgfβ2 signaling and reduction in phosphorylated-Smad2, causing reduced expression of Foxc1 and Pitx2, needed for proper development of the NC-derived periocular mesenchyme (Iwao et al., 2009).

12.5 DENTAL NC

During tooth development, the BMP-signaling and TGFβ-signaling transducers Smad1, Smad2, Smad3, Smad4, and Smad5 are strongly localized within the dental epithelium and cranial NC-derived dental mesenchyme, suggesting a critical role in regulating epithelial–mesenchyme interactions during tooth morphogenesis. The inhibitory Smad6 and Smad7 are expressed in a similar pattern, suggesting a requirement for negative feedback regulation of these pathways (Xu et al., 2003).

• • • •

CHAPTER 13

Sox Genes

The Sry-related HMG box (Sox) family consists of transcription factors that contain a conserved HMG box DNA-binding domain. Sox genes are often associated with regulation of fate specification and cellular differentiation during development and often function as transcriptional activators. These genes have been separated into classes from SoxA to SoxE based primarily on DNA-binding domain similarity (for phylogeny, see Bowles et al., 2000).

13.1 Sox2

Of the *SoxB* genes (*Sox1, 2, 3, 14, 19, and 21*), only *Sox2* has been reported to play a cell-autonomous role in NC development. *Sox2* is one of the early genes activated in the developing neural plate (Rogers et al., 2009), but it also has a broader role in NC lineages and in maintaining multipotency in other stem cell types. In *Xenopus*, induction of NC is always accompanied by neural plate induction, as marked by Sox2 (Bastidas et al., 2004). Although Sox2 is commonly used as a neural plate or early neural marker, neural plate expression of avian Sox2 is reduced as the NC segregates from the dorsal neural tube and migrating NCCs maintain a low Sox2 expression level. Expression of Sox2 in a subset of these cells that contribute to the developing PNS is subsequently upregulated and gradually becomes restricted to NC-derived glial sublineages (Aquino et al., 2006; Wakamatsu et al., 2004). Misexpression of Sox2 in embryonic ectoderm and in neural plate explants reveals that Sox2 inhibits NC formation, whereas later in the NC lineage (migratory and postmigratory NCCs, particularly those of the PNS), Sox2 plays a role in proliferation and differentiation (Wakamatsu et al., 2004). Conserved enhancers driving expression of chicken Sox2 in the NC have been identified (Uchikawa et al., 2003), and another study implicates the Nk-1-related gene Nbx as playing an upstream role in negatively regulating Sox2 expression and positively regulating the NC-inducing transcription factor Slug (Kurata and Ueno, 2003).

Sox2 is one of a select group of factors able to induce somatic cells to adopt a pluripotent state, as demonstrated in numerous induced pluripotent stem cell publications (Park et al., 2008; Takahashi and Yamanaka, 2006; Takahashi et al., 2007; Yu et al., 2009). Multiple NCSC lines express Sox2 (Aquino et al., 2006; Techawattanawisal et al., 2007; Thomas et al., 2008; Widera et al., 2009; Zhao et al., 2009). For instance, boundary cap NCSCs (bcNCSCs) are capable of differentiation into mature, functional Schwann cells (SCs) in the presence of neuregulins and express Sox2 in addition to other NC markers (Aquino et al., 2006). Multipotent NCSCs derived from rat periodontal ligament (PDL) and cultured as neurospheres express Twist, Slug, Sox2, and Sox9 (Techawattanawisal et al., 2007). A common theme recently uncovered by several studies of adult-derived NCSCs is the coexpression of a set of more NC-specific genes with a set of "reprogramming" genes, and Sox2 seems to be one factor that overlaps between the two groups. NCSCs capable of neuronal and glial differentiation derived from the rat palate (pNCSCs) and cultured as neurospheres express the NCSC markers Nestin, Sox2, and p75 as well as Klf4, Pou5f1/Oct4, and Myc and differentiated efficiently into neuronal and glial cells. This expression pattern seems to be conserved: the human palate expresses Nestin, Sox2, Pou5f1/Oct4, Klf4, and Myc (Widera et al., 2009). Multipotent skin-derived progenitors isolated from the pig (pSKPs), another adult-derived NCSC type, coexpress the pluripotency genes *Pou5f1/Oct4, Sox2, Nanog, and Stat3*, in addition to the NC markers p75, Twist1, Pax3, Snail2, Sox9, and Sox10, and these patterns likely apply to mouse and human SKPs (Zhao et al., 2009). Finally, self-renewing human NCC (hNCC) lines derived from pharyngulas have a transcript profile more similar to pluripotent ES cells than that of hNCC derivatives, and this includes the pluripotency markers Nanog, Pou5f1/Oct4, and Sox2 (Thomas et al., 2008).

13.2 Sox4

Of the SoxC genes (*Sox4, 11, 12, 22, and 24*), only *Sox4* has a demonstrated role in the NC. The expression pattern of Sox4 in endocardially derived tissue of the OFT and AV canal and tissues of NC origin such as the pharyngeal arches and craniofacial mesenchyme is similar between chicken and mouse embryos (Maschhoff et al., 2003; Ya et al., 1998). *Sox4*-null mouse embryos die at embryonic day 14, presumably due to cardiac failure. Observed heart defects are only in the arterial pole of the heart, suggesting that Sox4 is involved in interactions between NC-derived myofibroblasts and the endocardial components of the OFT (Ya et al., 1998). Consistent with this role, Sox4 and Nfatc transcription factors are associated with human forms of heart defects including rare truncus malformations (Restivo et al., 2006). Lithium exposure, a causative agent for certain teratogenic heart defects, reduces levels of both Sox4 and Nfats (Chen et al., 2008).

13.3 Sox5

Sox5 is the only member of the SoxD family (which includes Sox5, 6, 13, and 23) described in the NC. The long isoform of Sox5, LSox5, has a well-known role in chondrogenesis but, in addition to its expression in cartilage, is found in neuronal, glial, early NC, and other lineages, where it also functions. *Sox5*-null mice die neonatally due to respiration defects (Dy et al., 2008; Smits et al., 2001). During chondrogenesis, which includes the development of the NC-derived craniofacial skeleton, LSox5 coordinates with Sox6 and Sox9 to promote cartilage development by activating expression of cartilage-specific ECM molecules (Lefebvre et al., 1998; Perez-Alcala et al., 2004). Because Sox5 and Sox6 are highly similar, they have redundant function during chondrogenesis. Conditional deletion of floxed alleles of both Sox5 and Sox6 by Prrx1-Cre results in mice with chondrodysplasia (Dy et al., 2008), illustrating the importance of both of these SoxD members in cartilage formation.

In both chicken and mouse, LSox5 is also expressed in early NCCs; LSox5 expression initiates after the early NC markers Slug and Foxd3 and is maintained in the cranial glial lineage (Morales et al., 2007; Perez-Alcala et al., 2004; Stolt et al., 2008). LSox5 has NC-inducing properties; misexpression of LSox5 in the NT activates the migratory NC marker RhoB, and prolonged LSox5 expression expands the premigratory NC domain leading to overproduction of cranial NCCs (Perez-Alcala et al., 2004). In the chicken embryo, in addition to LSox5 expression in glia, a subpopulation of NC-derived differentiating neurons transiently expresses LSox5 (Morales et al., 2007). In the mouse embryo, Sox5 expression is observed in the melanocyte lineage but is not critical for melanocyte development; there is no observed melanocyte defect in *Sox5*-null mice (Stolt et al., 2008). However, the loss of LSox5 somewhat rescues the melanocyte defect of Sox10 heterozygous mutants. This genetic interaction led to the finding that Sox5 modulates Sox10 in the melanocyte lineage by recruiting the chromatin modifiers Ctbp2 and HDAC1 to the regulatory regions of target genes shared with Sox10, resulting in direct inhibition of Sox10-dependent activation (Stolt et al., 2008).

13.4 Sox8

All three members of the SoxE subgroup (Sox8, Sox9, and Sox10) are expressed and function in the NC and its derivatives, with some overlapping expression domains and functions (O'Donnell et al., 2006). Sox8 is highly similar to Sox9 and Sox10 and the protein has two separate transactivation regions. Sox8 is expressed in the NC, central nervous system, limbs, muscles, kidneys, adrenal glands, gonads, and craniofacial structures during mouse embryo development (Schepers et al., 2000; Sock et al., 2001). Homozygous null Sox8 mice have lower body weights than controls, but no other detectable defects in any of the Sox8-expressing lineages (Sock et al., 2001), and this is likely due to functional redundancy shared with Sox9 and Sox10. Two studies in particular demonstrate a genetic interaction between Sox8 and Sox10, suggesting that these highly similar proteins modulate expression or function of each other (Maka et al., 2005; Reiprich et al., 2008). In adrenal development, Sox8 (along with Sox10) is expressed in NCCs that migrate to the adrenal gland and is then downregulated in differentiating catecholaminergic adrenal cells. The NCCs of Sox10-deficient mice fail to colonize the developing adrenal medulla because of improper specification at the dorsal aorta and subsequent apoptosis during migration, and they do not express Sox8. Replacement of Sox10 with Sox8 significantly rescues the adrenal phenotype, suggesting functional redundancy. Sox8-null mice have only a minimal adrenal phenotype, but the phenotype in compound mutant Sox8-homozygous, Sox10-heterozygous embryos is much worse (Reiprich et al., 2008). These same compound mutants have an increased penetrance and severity of ENS defects (increased vagal NCC apoptosis and a resulting decrease in colonization of the gut), but Sox8 nulls have no ENS defects. This demonstrates that in ENS development, Sox8 functions as a modifier of Sox10. Like Sox10, Sox8 is also expressed in vagal and enteric NCCs and is later restricted to enteric glia (Maka et al., 2005), and these two SoxE factors may act together to maintain a pool of undifferentiated vagal NC stem cells (Maka et al., 2005). *Xenopus* Sox8, unlike in the mouse and chicken embryo, precedes Sox9 and Sox10 in NC progenitors. Sox8 expression persists in migrating cranial and trunk NCCs, in a pattern similar to Sox9 and Sox10. Sox8 knockdown in *Xenopus* does not inhibit NC formation but affects the timing of NC induction, impacting the later development of multiple NC lineages due to the inability of NCCs to migrate into the periphery (O'Donnell et al., 2006).

13.5 Sox9

13.5.1 Sox9 during Neural Crest Specification

Sox9 directs the development of NC, otic placodes, cartilage, and bone. In zebrafish, there are two Sox9 orthologs, Sox9a and Sox9b, which together perform the functions of the single-copy tetrapod Sox9 (Rau et al., 2006). Zebrafish Sox9b is expressed first in the prospective NC and then in cranial and trunk NC progenitors. Sox9b expression is extinguished in migrating NCCs, but some NC derivatives reactivate Sox9b expression at later stages (Li et al., 2002). *Xenopus* Sox9 is expressed in the prospective NC, where it has a role in regulating NC formation, and persists in migrating cranial NCCs as they populate the pharyngeal arches (Spokony et al., 2002). Knockdown of *Sox9* causes dramatic loss of NC progenitors, resulting in reduction or loss of the NC-derived craniofacial skeleton (Lee et al., 2004b; Spokony et al., 2002). Wnt- and BMP-regulated NC induction in *Xenopus* and chicken embryos depends on Sox9 transcriptional activator function (Lee et al., 2004b; Osorio et al., 2009; Sakai et al., 2006). Sox9 expression precedes expression of premigratory NC markers and establishes competence for NCCs to undergo an EMT, partly by directly activating the *Slug* promoter, but is not required for migration. Sox9 is, however, required for survival of trunk NCCs, which undergo apoptosis around the time of delamination if Sox9 is absent (Cheung et al., 2005; Sakai et al., 2006). Forced expression of Sox9 promotes NC-like properties in neural tube progenitors at the expense of CNS neuronal differentiation (Cheung and Briscoe, 2003). Sox9, together with Notch signaling, can induce ectopic Sox10 expression (Dutton et al., 2008).

13.5.2 The Role of Sox9 in Chondrogenesis

Sox9 is expressed in all cartilage progenitors, has an essential role in chondrogenesis, and marks commitment to chondrogenic differentiation (Sahar et al., 2005; Thomas et al., 1997). Heterozygous mutations in human *Sox9* result in campomelic dysplasia, a lethal disorder with skeletal malformations and craniofacial defects (Spokony et al., 2002). Sox9 regulates other early NC marker genes, the cartilage-specific gene Col2a1 and the bone-specific gene Runx2a (Akiyama et al., 2005). Col2a1 is directly activated by Sox9 and Sox10 and mediated by Sox9 and Sox10 cross-regulation and cAMP-dependent PKA signaling (Suzuki et al., 2006). Analysis of $Sox9^{Cre}$ knock-in mice demonstrates that Sox9 is expressed before Runx2, an early osteoblast marker gene, and that all osteochondrogenic cells are derived from Sox9-expressing progenitors (Akiyama et al., 2005). The zebrafish *Sox9a;Sox9b* double-mutant phenotype is additive; chondrocytes do not stack in Sox9a mutants, and in Sox9b, mutants do not expand properly, whereas compound mutants fail to do either, resulting in more severe craniofacial defects (Yan et al., 2002; Yan et al., 2005). Inactivation of Sox9 causes cranial NCCs to lose chondrogenic potential, converting to an osteoblast fate, marked by ectopic expression of osteogenic marker genes such as *Runx2*, *Osterix*, and *Col1a1* in the nasal cartilage region (Mori-Akiyama et al., 2003). Sox9 and Msx2 are coexpressed in a subpopulation of cranial NCC during migration that will form the mandible. Sox9 expression indicates chondrogenic lineage determination, but Msx2 represses chondrogenic differentiation until cranial NCC migration is completed (Takahashi et al., 2001). Bmp4 induces chondrogenesis at sites where Sox9 expression is high relative to that of Msx2, and ectopic chondrogenesis is associated with Sox9 and Msx2 upregulation adjacent to Bmp4 signals (Semba et al., 2000). During cranial suture closure, Sox9 is upregulated along with cartilage markers, and haploinsufficiency of Sox9 results in delayed suture closure (Sahar et al., 2005).

13.5.3 Sox9 in NCSCs

In the chicken embryo, high levels of Sox9 (or Sox10 or Sox8) expression in the NT induce a migratory NC-like phenotype and maintains these cells in an undifferentiated state (McKeown et al., 2005). Sox9 is one of the early NC transcription factors often expressed in multipotent NC-derived progenitors cells. Sox9 expression is maintained in multipotent NCSCs derived from rat PDLs (Techawattanawisal et al., 2007), multipotent dental NC-derived progenitor cells (Degistirici et al., 2008), and multipotent skin-derived precursors (Zhao et al., 2009). Sox9 is also overexpressed with another NCSC marker, Twist1, in NC-derived malignant peripheral nerve sheath tumors (MPNSTs) (Miller et al., 2006).

13.5.4 Sox9 in the Melanocyte Lineage

Sox9 is expressed by melanocytes in neonatal and adult human skin and activates transcription of *Mitf*, *Dct*, and *Tyr*. Within the melanocyte lineage, Sox9 is upregulated by cAMP and PKA signaling in response to UVB exposure and is downregulated by the secreted factor Agouti signal protein, resulting in decreased pigmentation (Passeron et al., 2007).

13.6 Sox10

Sox10 is an extremely important regulator at multiple steps of NC development and has been implicated in interactions with many other transcription factors in the exquisite control of multiple cell fates (Figure 13.1).

13.6.1 Sox10 Expression during NC Induction and in Migratory NCCs

During NC development, Sox10 is first expressed in the prospective NC in the dorsal NT, and continues in multipotent migratory NCCs (Ishii et al., 2005; Young et al., 2004), in NC-derived ENS and PNS progenitors, and in the melanocyte lineage (Nonaka et al., 2008). In *Xenopus*, Sox10 is expressed in prospective NC and otic placode regions from the earliest stages of NC specification in a Wnt and FGF-dependent manner, and in migrating cranial and trunk NCCs (Aoki et al., 2003;

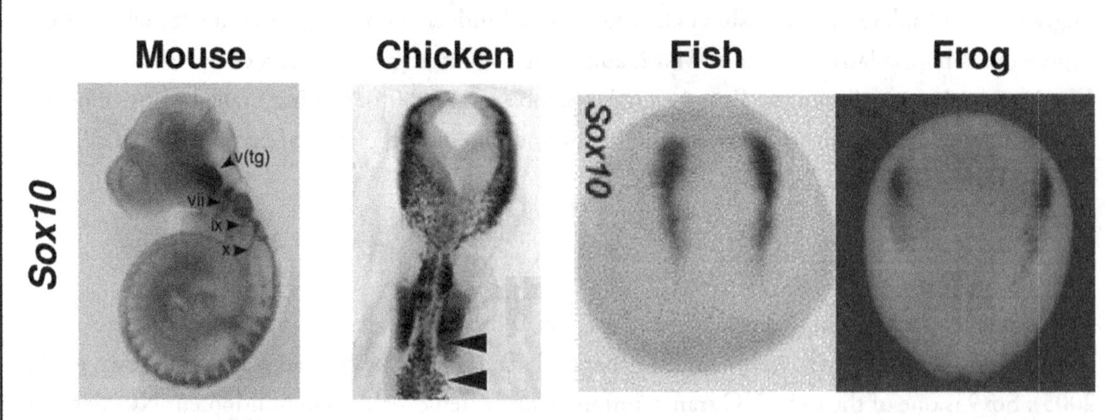

FIGURE 13.1: Whole-mount in situ hybridizations show *Sox10* in mouse, chicken, fish, and frog embryos. Mouse: Bennetts et al., 2007, 9.5-dpc embryo, lateral view; chicken: Basch et al., 2006, stage 8 embryo, dorsal view; zebrafish: Rau et al., 2006, three-somite embryo, dorsal view; *Xenopus*: O'Donnell et al., 2006, stage 13/14 embryo, dorsal view.

Honore et al., 2003). Sox10 is initially expressed in all NCCs, colocalizing with Slug and Sox9, major regulators of NC formation, but is then downregulated in the cranial NC and persists mostly in NCCs from the trunk region (Aoki et al., 2003). In the chicken embryo, Sox10 is expressed in migrating NCCs just after Slug but is lost as cells undergo differentiation in ganglia of the PNS and ENS. It is also expressed in the developing otic vesicle, the developing CNS, and the pineal gland (Cheng et al., 2000). In the mouse, where a *Sox10βGeoBAC* transgene closely approximates Sox10 endogenous expression, Sox10-driven *lacZ* expression can be detected in the anterior dorsal NT at E8.5 and in cranial ganglia, otic vesicle, developing DRGs, thyroid parafollicular cells, thymus, and salivary, adrenal, and lacrimal glands (Deal et al., 2006; Muller et al., 2008). Sox10 expression is generally present in migratory NCCs that will give rise to nonmesenchymal NC derivatives, such as melanocytes, glia, and neurons. This includes ANS neurons such as those contributing to the lung ganglia and NC-derived pancreatic innervation (Burns and Delalande, 2005; Nekrep et al., 2008). Sox10-expressing cells have been fate-mapped by vital dye injections in *Sox10::eGFP* transgenic zebrafish (Wada et al., 2005), showing that some Sox10-expressing cells contribute progeny to the paired trabeculae and ethmoid bones (in zebrafish) (Wada et al., 2005).

Sox10 functions at many stages of NC development: NC formation; maintenance of multipotent migratory NCCs; and survival, specification, and differentiation of nonmesenchymal NC derivatives including melanophores and peripheral or enteric glia (Drerup et al., 2009; Elworthy et al., 2003; Kelsh, 2006; Kim et al., 2003). The effect of Sox10 loss on later stages of NC derivatives will be discussed later in this chapter (see Section 13.6.4). In *Xenopus*, Sox10 knockdown causes loss of NC precursors and Slug and Foxd3 expression, enlargement of non-NC domains, and increased apoptosis and decreased proliferation (Honore et al., 2003). Knockdown of Sox10 blocks induction of melanocytes and ganglia in vivo and in vitro (Honore et al., 2003). Overexpression of Sox10 causes a large expansion of the Slug expression domain in *Xenopus* and induces expression of HNK-1 in a broad neuroepithelial domain in the chicken embryo (Aoki et al., 2003; McKeown et al., 2005). *Xenopus* Sox10-injected embryos show a later increase in Trp2-expressing pigment cells, suggesting a role for Sox10 in melanocyte lineage specification (Aoki et al., 2003). Many more cells emigrate from the NT in chicken embryos with Sox10 overexpression, but these cells never express differentiated markers, indicating Sox10 maintains an undifferentiated state or inhibits terminal differentiation (McKeown et al., 2005). Addition of Sox10 to cultured cells inhibits neuronal and glial differentiation of multilineage ENS progenitor cells without interfering with neurogenic commitment (Bondurand et al., 2006).

Sox10 and Sox8 are expressed in NCCs migrating to the adrenal gland and downregulated in catecholaminergic cells (Gut et al., 2005; Reiprich et al., 2008). Sox10-deficient NCCs are not properly specified; they undergo apoptosis and do not colonize the adrenal anlage, resulting in loss of the adrenal medulla. Sox10 and Sox8 may share functional redundancy, as *Sox8* homozygous-null mutations alone have only minimal effect on adrenal gland development, but defects are seen in *Sox8* homozygous-null, *Sox10* heterozygous-null compound mutants, and extra Sox8 can partially rescue *Sox10*-null adrenal defects. Also, Sox8 expression is regulated by Sox10; *Sox10*-null NCCs do not express Sox8 (Reiprich et al., 2008). Sox10 and Sox9 regulate each other and directly activate Col2a1 expression during cartilage and NC differentiation, and activation is further enhanced by cAMP-dependent PKA signaling (Suzuki et al., 2006). In homozygous *Erbb3*^{msp1} embryos, *Sox10*^{LacZ} expression is absent in cranial ganglia and sympathetic chains, but development of other Sox10-expressing cells appears unaffected (Buac et al., 2008). Sox10 activates Krox20 by binding an NC-specific enhancer in the *Krox20* locus in synergy with Krox20 itself. Inactivation of Sox10 prevents maintenance of Krox20 expression in the migrating NC (Ghislain et al., 2003).

13.6.2 Transcription Factors Upstream of Sox10 in NC Development

In chicken embryos, Pax7 is required for NC formation in vivo; blocking its translation inhibits expression of NC markers including Sox10 (Basch et al., 2006). Sox9 regulates other early NC genes including *Sox10* (Yan et al., 2005). The zebrafish *Foxd3*$^{mother\ superior}$ mutation (which causes loss of Foxd3 NC expression) leads to a depletion of NC derivatives preceded by reduction in NC-expressed transcription factors including Sox10 (Montero-Balaguer et al., 2006). Zebrafish *Foxd3*^{sym1} (a putative null allele) mutants have normal numbers of premigratory NCCs but reduced levels of Sox10 and Snail (Stewart et al., 2006). Snail, but not Slug, controls Sox10 expression in *Xenopus* (Honore et al., 2003). Zic4 is involved in induction of NC markers Sox10 and Slug (Fujimi et al., 2006). *Disc1* mutant cranial NCC migration defects correlate with enhanced expression of Foxd3 and Sox10, leading to a loss of craniofacial cartilage and expansion of peripheral cranial glia. Disc1 functions in transcriptional repression of Foxd3 and Sox10, mediating cranial NCC migration and differentiation (Drerup et al., 2009).

13.6.3 Regulation of Sox10

Three clusters of highly conserved sequences in the *Sox10* promoter, one of which shows strong enhancer potential in cultured melanocytes, are disrupted in the transgene-insertion mutant mouse line (*Sox10*^{Hry}) that has a large deletion in a distal upstream enhancer region in the *Sox10* locus, resulting in decreased Sox10 expression, aganglionosis, and melanocyte loss in homozygous mutants (Antonellis et al., 2006). Multiple conserved enhancers of *Sox10* containing Sox, Pax, AP-2, and Tcf/Lef binding sites show distinct but overlapping activities including expression in several NC derivatives such as the developing PNS, Schwann cells, melanocytes, and adrenal gland (Antonellis et al., 2008; Werner et al., 2007). Some enhancers seem to direct pan-NC regulatory control (Antonellis et al., 2008). Characterization of *Sox10βGeoBAC* expression confirms presence of essential regulatory regions for the PNS lineage (Deal et al., 2006). The 3' end of the conserved first intron is required for proper spatial expression of Sox10 and contains conserved binding sites for transcription factors known to be essential in early NC induction, including Tcf/Lef, Sox and Foxd3; and β-catenin, Notch signaling, and Sox9 can all induce ectopic Sox10 expression in early embryos (Dutton et al., 2008).

The activities of individual SoxE factors are regulated by sumoylation, a posttranslational modification affecting protein stability, activity, and localization. Sumoylated forms mediate some specific activities, whereas nonsumoylated forms mediate a separate set of activities (Taylor and Labonne, 2005). Sox10 protein has three sumoylation consensus sites that modulate Sox10 activity. Sox10 sumoylation represses its transcriptional activity on the target genes *Gjb1* and *Mitf* by modulating its synergistic interaction with cofactors Krox20 and Pax3 at these promoters (Girard and Goossens, 2006). Sox10 contains two nuclear localization signals and is most frequently detected in the nucleus, but Sox10 is an active nucleocytoplasmic shuttle protein. Sox10 has a functional Rev-type nuclear export signal (NES) within its DNA-binding domain. Mutational inactivation of this NES or treatment of cells with the CRM1-specific export inhibitor leptomycin B inhibited shuttling of Sox10 from the nucleus to the cytoplasm. Inhibition of Sox10 nuclear export decreased transactivation of transfected reporters and endogenous target genes (Rehberg et al., 2002). Consistent with a role as a shuttling transcription factor, Sox10 interacts with Armcx3, an integral membrane protein of the mitochondrial outer membrane containing an armadillo repeat, and consequently is associated with the mitochondrial outer membrane when it is in the cytoplasm. Armcx3 does not possess transcriptional activity but enhances Sox10 transactivation of *nAChR α3* and *β4* subunit genes (Liu et al., 1999; Mou et al., 2009). Many target gene promoters have multiple Sox10-binding sites. Sox10 can bind target DNA as a monomer or dimer. Dimers generally bind through two heptameric binding sites in a specific orientation and spacing, mediated by an N-terminal DNA-dependent Sox10 dimerization domain. Dimers have a higher binding affinity and increase the angle of DNA bending (Peirano and Wegner, 2000). Specific amino acid residues in a conserved region immediately preceding the HMG domain of Sox10 are required for cooperative binding with the HMG domain during dimeric binding. Maintenance of cooperativity is essential for full activation of target promoters such as the *myelin protein zero* (*Mpz*) gene, but dimer-dependent conformational changes such as bending angle introduced into the promoter are less important (Schlierf et al., 2002). The Sox10 dimerization and transactivation domains are needed in melanocyte and ENS development, but not in early NC development. The transactivation domain is required for satellite glia differentiation and Schwann cell myelination, whereas the DNA-dependent dimerization domain is required for the transition of immature Schwann cells to the promyelinating stage (Schreiner et al., 2007).

13.6.4 Sox10 Mutants as a Model for Waardenburg Syndrome 4

Waardenburg syndrome type 4 (WS4), also known as Waardenburg–Shah or Hirschsprung–Waardenburg syndromes, encompasses a range of phenotypes with characteristics of both WS (pigment deficiencies and deafness) and Hirschsprung disease (HSCR) (aganglionic megacolon). Haploinsufficiency of Sox10 causes pigmentation and megacolon defects, also observed in *Sox10*$^{Dom/+}$ mice and WS4 patients with heterozygous SOX10 mutations (Britsch et al., 2001). The white spotting and aganglionic megacolon of *Sox10*^{Dom}, *Ednrb*$^{piebald\text{-}lethal\ (sl)}$, and *Edn3*$^{lethal\ spotting\ (ls)}$ mouse mutants are similar to the WS4 phenotype (Pingault et al., 1998). *Sox10*$^{Dom/+}$ mice have pigmentation deficiency or dysganglionosis 93% of the time (dysganglionosis 79% and pigmentation deficiency 90%), and both defects are seen in 68% of mice (Brizzolara et al., 2004). Mutations in the *Ret* gene are responsible for approximately half of familial HSCR cases and some sporadic cases. Mutations in genes encoding Ret ligands (*Gdnf* and *Ntn*), components of the Endothelin signaling pathway (*Ednrb*, *Edn3*, *Ece1*), and *Zeb1* and *Sox10* have also been identified in patients with HSCR (Benailly et al., 2003; Parisi and Kapur, 2000). WS4 patients have mutations in *Sox10*, but *Sox10* mutations are usually not detected in patients with HSCR alone (Pingault et al., 1998). Haploinsufficiency of Sox10 is most often associated with disorders beyond HSCR. Human Sox10 is essentially expressed in NC derivatives that form the PNS, and in the adult CNS, but is more widely expressed than in rodents, with weak transcriptional activator activity (Bondurand et al., 1998; Kuhlbrodt et al., 1998b). Because of genetic background differences, unknown modifiers, and different types of Sox10 lesions, the human WS4 phenotype among patients with *Sox10* mutations covers a wide range, from chronic intestinal pseudoobstruction alone to classic WS4 to WS4 with severe peripheral neuropathies like Charcot–Marie–Tooth neuropathy type 1 (CMT1) or PCWH (peripheral demyelinating neuropathy, central dysmyelinating leukodystrophy, WS, and HSCR), and dysfunction of Sox10 may extend to the CNS, resulting in brain phenotypes (Inoue et al., 1999; Pingault et al., 2000; Southard-Smith et al., 1999; Sznajer et al., 2008; Touraine et al., 2000; Verheij et al., 2006).

Sox10 plays a cell-autonomous role in NCCs migrating into the GI tract and is essential for proper ENS development. *Sox10*$^{Dom/+}$ mice have deficiencies of NC-derived enteric ganglia in the distal colon, whereas *Sox10*$^{Dom/Dom}$ embryos have total enteric aganglionosis and die late in embryogenesis or perinatally (Herbarth et al., 1998; Southard-Smith et al., 1998). In *Sox10*$^{Dom/Dom}$ embryos, apoptosis was increased in sites of early NCC development before these cells enter the gut (Herbarth et al., 1998; Kapur, 1999). Cell death is also increased before lineage segregation in undifferentiated, postmigratory NCCs lacking Sox10. Surviving *Sox10*-null NCCs do not adopt a glial fate, even in gliogenic conditions. In *Sox10* heterozygous mutant NCCs, survival is normal, but fate specification is still drastically affected (Paratore et al., 2001; Paratore et al., 2002b). Mutant enteric NC-derived cells do not maintain a progenitor state and acquire preganglia traits, resulting in a reduction of the progenitor pool size (Paratore et al., 2002b).

Zebrafish *Sox10*^{colourless} mutant NCCs, another model of WS4, form only mesenchymal NC fates, whereas NCCs that would normally adopt nonmesenchymal NC fates fail to migrate and differentiate and instead undergo apoptosis. Mitf expression is disrupted in *Sox10*^{colourless} mutants, giving rise to melanophore defects (Dutton et al., 2001). ENS progenitor fate specification, marked by Phox2b, is defective in these mutants, with most NCCs failing to migrate to the GI tract primordium (Elworthy et al., 2005).

In addition to *Sox10*, mutations in *Edn3* and *Ednrb* have been associated with WS4 (Karaman and Aliagaoglu, 2006; Matsushima et al., 2002). The *Ednrb* ENS enhancer has Sox10 binding sites, and Sox10 and Ednrb interact genetically. *Sox10;Ednrb* (and *Sox10;Edn3*) compound mutants have a drastic increase in white spotting, absence of melanocytes within the inner ear, and more severe ENS defects. In the GI tract of double heterozygous mutants, no apoptosis, proliferation, or differentiation defects in NCCs were detected, but apoptosis was increased in vagal NCCs outside of the GI tract (Cantrell et al., 2004; Stanchina et al., 2006). Sox10-expressing enteric NCC progenitors are reduced in *Edn3*-deficient embryos, suggesting endothelin signaling is necessary for maintenance and proliferation of Sox10-expressing ENS progenitors (Bondurand et al., 2006). Pax3, required for normal enteric ganglia formation, functions with Sox10 to activate transcription of *Ret* (Lang et al., 2000). Sox10 expression in enteric NCCs is decreased in *Pofut1*-null embryos that have defective Notch signaling, whereas enteric NCCs expressing Ascl1, a strong repressor of Sox10 suppressed by Notch signaling, are increased. Notch signaling is required for maintenance of ENS progenitors and Sox10 expression by attenuating a cell-autonomous neuronal differentiation program, at least in part by suppressing *Ascl1* (Okamura and Saga, 2008). Sox8 is expressed with Sox10 in migrating vagal and enteric NCCs and is later confined to enteric glia. Loss of Sox8 alone had no effect on the ENS, but loss of *Sox8* alleles in *Sox10* heterozygous mice increased the penetrance and severity of *Sox10* ENS defects and impaired early colonization of the GI tract by enteric NCCs, drastically increasing apoptosis in vagal NCCs outside the GI tract. The defects in ENS development of mice with *Sox10* and *Sox8* mutations may be caused by a reduction of the pool of undifferentiated vagal NCCs (Maka et al., 2005). Binding sites for Sox10 exist in the *Ednrb* ENS enhancer, suggesting that Sox10 may have multiple roles in regulating *Ednrb* in the ENS (Zhu et al., 2004).

13.6.5 Sox10 in Glia of the PNS

Sox10 first appears in developing NC and continues as NCCs contribute to the developing PNS, where it functions in satellite glia and Schwann cells and is a marker of Schwann cell differentiation (Kuhlbrodt et al., 1998a; Miller et al., 2006; Schreiner et al., 2007). Sox10 is also expressed in the CNS, first in glial precursors and later in myelin-forming oligodendrocytes of the adult brain (Kuhlbrodt et al., 1998a; Stolt et al., 2002). In the PNS, Sox10 is a key regulator of specification and differentiation of peripheral glial cells (Schwann cells) (Britsch et al., 2001; Svaren and Meijer, 2008). In *Sox10*-null mutants, neuronal cells form in DRGs, but Schwann cells and satellite glia are lost, resulting in later degeneration of sensory and motor neurons (Britsch et al., 2001; Stolt et al., 2002). During Schwann cell development, Sox10 functions synergistically with Oct6, and modulates the functions of Pax3 and Krox20, conferring specificity to these transcription factors in developing and mature glia (Kuhlbrodt et al., 1998a). Sox10 partners synergistically with NFATc4 to activate Krox20, regulating myelination genes (Kao et al., 2009). In glial cells, Sox10 associates with the N-myc interactor (Nmi) protein, which modulates Sox10 transcriptional activity in reporter assays, perhaps in a promoter-specific manner (Schlierf et al., 2005). Among the primary targets, Sox10 controls expression of Erbb3, a Neuregulin receptor, in NCCs. Downregulation of Erbb3 accounts for many defects seen in mutants, but Sox10 has functions not mediated by Erbb3, such as in the melanocyte lineage (Britsch et al., 2001). Sox10 directly regulates *myelin protein zero* (*Mpz* or *P0*) in Schwann cells (Peirano and Wegner, 2000). Other direct target genes of Sox10 are the *myelin proteolipid protein* (*Plp1*), *extracellular superoxide dismutase* (*Sod3*), and *pleiotrophin* (*Ptn*), and *Sox10* itself (Lee et al., 2008b).

13.6.6 Sox10 in Neural Derivatives

In migrating multipotent NCCs, Sox10 preserves glial and neuronal potential. Sox10 is needed in vivo for induction of the neurogenic factors Ascl1 and Phox2b and inhibition or delay of neuronal differentiation at higher dosages (Kim et al., 2003). Undifferentiated enteric NCCs express Sox10, Phox2b, p75, and Ret. At E10.5, between 10% and 15% of these NC-derived cells in the small intestine have started to differentiate into neurons. By E12.5, 25% of Phox2b-expressing cells in the small intestine express neuronal markers such as ubiquitin carboxy-terminal hydrolase (PGP9.5), and this fraction increases to 47% at E14.5 (Young et al., 2003). Although Sox10 is maintained in glial precursors along with p75 and Fabp7 and low Ret expression, differentiating enteric neurons no longer express Sox10 and have low p75, but high Ret, expression (Young et al., 2003). Sox10 expression in undifferentiated multipotent NCCs is mutually exclusive to Foxs1, which marks emerging DRG sensory neurons. Foxs1-negative, Sox10-positive migrating NCCs with a high proliferative activity surround Foxs1-positive, Sox10-negative pioneering NCC neuronal progenitors with limited proliferation (Montelius et al., 2007). These data suggest that Sox10 may play an early role in maintaining neuronal differentiation potential of NCCs.

Sox10 mutants have a complete absence of glial differentiation but apparently normal initial neurogenesis. However, without glial cells, motor neurons and sensory neurons degenerate later in development. As neuronal and glial precursors are generated and segregated from NCCs in the DRGs, these NCCs have increased apoptosis and decreased proliferation (Sonnenberg-Riethmacher et al., 2001). In addition to affecting progenitors, there is some evidence that Sox10 may play a direct role in NC-derived sensory neurons, but this is not easy to demonstrate in mammalian embryos because of interdependence of closely associated neurons and glia. However, in zebrafish, early DRG sensory neuron survival may be independent of glia. Sox10 is expressed transiently in the sensory neuron lineage and specifies sensory neuron precursors by regulating *Neurog1* (Carney et al., 2006). In zebrafish, glial cells and their target axons coalesce at an early stage and are coupled throughout migration, with axons providing instructive cues necessary to direct glial migration. Genetic ablation of glia in Sox10 mutants, uncoupling axon and glial migration, shows Sox10 has an important role in nerve fasciculation (Gilmour et al., 2002). Inactivation of zebrafish Sox10 or Neurog1 leads to more than twice the normal number of neuromasts along the posterior lateral line. Development of intercalary extra neuromasts may occur because of the absence of NC-derived peripheral glia, which may inhibit the assembly of interneuromast cells into neuromasts (Lopez-Schier and Hudspeth, 2005).

13.6.7 Sox10 in Melanocytes and Regulation of Mitf

As evident from analysis of *Sox10*^{Dom} mice and WS4 phenotypes, Sox10 is a critical factor for melanocyte development. *Sox10*$^{Dom/Dom}$ embryos lack NC-derived cells expressing Mitf, Dct, and Kit, and NCC primary cultures from these embryos do not give rise to pigmented cells. In Sox10$^{Dom/+}$ heterozygous embryos, melanoblasts expressing Kit and Mitf are present in reduced numbers, and pigmented cells eventually develop in nearly normal numbers both in vitro and in vivo (Potterf et al., 2001). Sox10 directly transactivates the melanocyte "master regulator" gene *Mitf* 100-fold through a conserved binding site and this transactivation is further stimulated by Pax3 (Potterf et al., 2000), much like the synergistic activation by Sox10 and Pax3 on the *Ret* promoter (Lang and Epstein, 2003). In the context of *Ret*, Sox10 mutants that cannot bind DNA retain the ability to activate the enhancer in the presence of Pax3, but in the context of *Mitf*, Pax3 and Sox10 must each bind independently to the DNA (Lang and Epstein, 2003). A Sox10 mutant with a C-terminal truncation acting as a dominant-negative reduces *Mitf* induction (Potterf et al., 2000). *Mitf* is also activated by cAMP signaling through Creb1, but the direct Creb1 activation of *Mitf* requires Sox10 to bind a second DNA element. In melanoma and neuroblastoma cells, activation of *Mitf* by either Sox10 or cAMP is mutually dependent, and ectopic Sox10 with cAMP signaling is sufficient to activate the *Mitf* promoter in neuroblastoma cells (Huber et al., 2003). *Sox10*-null zebrafish mutants also lack expression of Mitfa, the equivalent of Mitf-M. Reintroduction of *Mitfa* expression in NCCs can rescue melanophore development in *Sox10*-null zebrafish embryos, suggesting that the essential function of Sox10 during melanophore development is activation of *Mitfa* (Elworthy et al., 2003). Sox10 and Mitf-M are also expressed in melanoblasts migrating toward the prospective inner ear of mouse embryos but are later separately expressed in different cell types of the newborn cochlea (Watanabe et al., 2002a).

13.6.8 Coregulation of Mitf Target Genes by Sox10

Mitf controls a set of genes critical for pigment cell development and pigmentation, including *Dct*, *Tyr*, and *Tyrp1*, but Sox10 also augments activation of these genes (Passeron et al., 2007). The *Tyr*,

Tyrp1, and *Dct* promoters all contain an E-box bound by Mitf and binding sites for RPE-specific factors such as Otx2 or for melanocyte-specific factors such as Sox10 or Pax3 (Murisier and Beermann, 2006). Sox10 colocalizes with Dct in early melanoblasts before Tyr or Tyrp1 expression, $Sox10^{Dom/+}$ melanoblasts have a transient loss of Dct expression, and Sox10 transactivates the Dct promoter in vitro (Jiao et al., 2004; Potterf et al., 2001). Critical melanocyte-specific enhancers in the *Dct* promoter contain Sox10 and Mitf binding sites. Sox10 and Mitf on their own directly activate *Dct* transcription, but together activate *Dct* expression synergistically (Jiao et al., 2004; Jiao et al., 2006). Mitf in melanoma cells is modified by sumoylation, and an Mitf substitution mutation affecting sumoylation has enhanced synergy with Sox10 on the *Dct* promoter (Murakami and Arnheiter, 2005). A BAC transgene containing the *Tyrp1* gene and surrounding sequences that recapitulates endogenous expression has a conserved melanocyte-specific enhancer activated by Sox10 (Murisier et al., 2006).

13.6.9 Additional Sox10 Regulation

Sox10 activity in the melanocyte lineage may be modulated by Sox5, which competes with Sox10 for shared target binding sites and recruits the corepressors Ctbp2 and Hdac1 to inhibit Sox10 target genes. Loss of Sox5 partially rescues reduced melanoblast generation and gene expression in *Sox10* heterozygous mice (Stolt et al., 2008). $Gli3^{Mos1}$, a truncation mutation identified as a modifier that increases the severity of $Sox10^{LacZ/+}$ defects, causes more drastic reduction of pigmentation when present as a heterozygous allele in combination with $Sox10^{LacZ/+}$. $Gli3^{Mos1/Mos1}$ embryos have reduced transcripts of Sox10 and early melanoblast markers Mitf, Dct, and Si, suggesting disrupted melanoblast specification (Matera et al., 2008). Ednrb is expressed with Sox10 in melanoblasts. *Ednrb*-null mice maintain Sox10 expression in melanoblasts; unlike in the ENS, Sox10 does not directly activate *Ednrb* transcription in the melanocyte lineage, suggesting context-dependent regulation by an unknown mechanism (Hakami et al., 2006). In zebrafish, Sox10 and leukocyte tyrosine kinase (Ltk) are required for iridophore specification from NCC, and like *Sox10* mutants, zebrafish Ltk^{shady} mutants lack iridophores. *Sox10* heterozygous mutants have an increase in Ltk-expressing cells, but cells also retaining Sox10 expression do not express other NCSC markers and may represent lineage-restricted progenitors or incompletely specified multipotent progenitors (Lopes et al., 2008).

13.6.10 Sox10 in Melanocyte Stem Cells

The bulge region of the adult hair follicle contains a niche for melanocyte stem cells. Development of melanocyte stem cells is controlled by Pax3, Sox10, and Mitf, and extracellular cues such as Wnt (Choi et al., 2008; Sommer, 2005). Many white spotting genes (*Sox10*, *Pax3*, *Mitf*, *Slug*, *Ednrb*, *Edn3*, *Kit*, *Kitl*) are associated with hypopigmentary disorders and deafness as a result of NCSC-derived melanocyte deficiency (Hou and Pavan, 2008). Graying hair is typically caused by defective cell migration into the bulb of the hair, and reduction of Sox10, Pax3, Mitf, and their target genes *Tyr* and *Tyrp1* (Choi et al., 2008).

13.6.11 Sox10 in NCSCs and Cancer

Sox10 is one of the NCSC markers expressed in several different adult-derived NCSC types. Multipotent stem cell-like SKPs from facial skin of adult mice, pigs, and humans express Sox10 and p75 in addition to pluripotency-related genes (Wong et al., 2006; Zhao et al., 2009). A boundary cap is a structure composed of late migrating NC at the dorsal root entry zone and motor neuron exit points. It contains multipotent stem cells and can generate neurons and peripheral glia during embryogenesis. Sox10 marks multipotent progenitors in boundary caps and in cultured bcNCSCs (Aldskogius et al., 2009; Aquino et al., 2006).

Genetic programs essential to NC development are often activated in neuroectodermal tumors, which express Sox10, AP-2α, and Pax3. Both Sox10 and AP-2α are generally expressed in relatively differentiated neoplasms (Gershon et al., 2005). Sox10 is expressed in some neuroblastoma cell line subtypes (Acosta et al., 2009), almost all melanomas and about half of all MPNSTs, and diffusely in some schwannomas and neurofibromas (Nonaka et al., 2008). Sox10 and its target Erbb3 are both overexpressed in pilocytic astrocytoma relative to other pediatric brain tumors (Addo-Yobo et al., 2006).

• • • •

CHAPTER 14

Zinc Finger Genes

14.1 Gata2/3

The Gata family consists of zinc finger transcription factors that bind the sequence G-A-T-A. In the chicken embryo, Gata2 (Gata3 in the mouse) is an essential member of the transcription factor network controlling sympathetic neuron development, which includes Ascl1, Hand2, Phox2a, and Phox2b. Gata2 is expressed in developing sympathetic neuronal precursors, beginning after Ascl1, Phox2b, Hand2, and Phox2a but before the noradrenergic marker genes Th and Dbh (Tsarovina et al., 2004). Gata2 expression increases in response to the other factors in this network, which are all BMP-responsive factors, and is itself responsive to BMP signaling. As further evidence that Gata2 is downstream of these factors, Gata2 is not expressed in *Phox2b*-null mice (Tsarovina et al., 2004). Gata2 loss-of-function decreases sympathetic chain size and reduces Th expression. Ectopic expression of Gata2 in chick NC precursors results in neuronal differentiation toward a nonautonomic, non-Th-expressing phenotype, demonstrating that it is sufficient for autonomic neuron differentiation but requires coregulators within the sympathetic lineage (Tsarovina et al., 2004). In zebrafish *Hand2*$^{\textit{hands off}}$ mutants, sympathetic ganglion primordia expressing Phox2a, Phox2b, and Ascl1 are formed, but these cells have significantly reduced expression of Gata2 (Lucas et al., 2006).

Mammalian Gata3 regulates development of the thymus and adrenal glands, SNS, ear, and kidney, and is implicated in the formation of autosomal dominant hypoparathyroidism, sensorineural deafness, and renal anomaly syndrome (Airik et al., 2005; Raid et al., 2009). Expression of Gata3 causes an increase in Dbh- and Th-expressing neurons in primary NCSC culture. Gata3 transactivates the norepinephrine-synthesizing *Dbh* gene promoter together with Sp1 and AP4 factors bound to the *Dbh* promoter (Hong et al., 2008b). Gata3 transactivates the *Th* promoter via a promoter element that contains a binding site for Creb, but not Gata3, which physically interacts with Creb both in vitro and in vivo (Hong et al., 2006). A 625-kb *Gata3* YAC transgene is missing regulatory elements that confer Gata3 expression in a subset of NC-derived lineages (thymus and sympathoadrenal system) and fails to rescue embryonic lethality, suggesting that neuroendocrine deficiency in the SNS causes death (Lakshmanan et al., 1999). *Gata3*-null mice have reduced Th and Dbh leading to a reduction in norepinephrine, but several other SNS genes are not altered (Lim et al., 2000).

In *Gata3*-null embryos, norepinephrine deficiency is the immediate cause of embryonic lethality, usually between E11 and E12 (Hong et al., 2006). Treatment with sympathomimetic beta-adrenergic receptor agonist or catechol intermediates (Lim et al., 2000; Raid et al., 2009) improves survival up to E18, allowing for further analysis of heart defects. These rescued mutants have heart defects such as ventricular septal defects, double outlet right ventricle, aortic arch abnormalities, and persistent truncus arteriosus. Short OFT and insufficient rotation of the truncus arteriosus during looping may be the main causes of these malformations (Lim et al., 2000; Raid et al., 2009). A *Gata3^{lacZ}* knock-in recapitulated the endogenous Gata3 and was robustly expressed in the endocardial ridges and endothelium of distal OFT. Reporter expression was also strong in the mesenchyme and epithelium of ventral pharyngeal arches and lower in the AV canal (Raid et al., 2009). In the ventricles and atria, Gata3 is not expressed after E13.5, but LacZ activity persisted, allowing lineage tracing of cells formerly expressing Gata3,with activity detected in semilunar valves, atrioventricular valves, OFT, and aortic arch arteries (Raid et al., 2009). Gata2/3 may be important for expression of key cardiovascular development genes like *Hand2* (Ruest et al., 2004). The *Hand2* promoter has two conserved Gata binding sites required for *Hand2* expression within the right ventricle, suggesting a role for direct regulation by Gata proteins. Gata factors are not restricted to the right ventricle, so they are likely to cooperate with coregulators to achieve right ventricular-specific regulation (McFadden et al., 2000). Cardiac-specific response to Edn1 requires synergy between the mostly ubiquitous SRF and tissue-restricted Gata proteins, which bind an Edn1 response element. SRF and Gata proteins form a complex and in transient cotransfections synergistically activate other Edn1-inducible promoters that contain both Gata and SRF binding sites (Morin et al., 2001).

14.2 Gata4

Gata4 is expressed in many NCSCs. Most migratory NCCs in E9.5–E11.5 embryos are labeled by a transgenic 5-kb Gata4 promoter driving expression of a fluorescent reporter (Pilon et al., 2008). Between E10.5 and E11.5, the Gata4 reporter is more active in boundary cap cells, where the presence of cells with NCSC properties has been corroborated (Aquino et al., 2006). In vitro analysis of the properties of Gata4-expressing cells also supports the expression of Gata4 within NCSCs (Pilon et al., 2008).

14.3 Gata6

Gata6 is strongly expressed in embryonic ectoderm, NT, and NC-derived cells, closely parallel to expression of Bmp4, a direct downstream target of Gata4 and Gata6 (Nemer and Nemer, 2003). Gata6 protein is also abundant in parts of the gut, pulmonary system, myocardium of the heart, and regions of NC- and sclerotome-derived chondrogenesis, but significantly reduced within the endocardial cushions and OFT of the heart, regions expressing the highest levels of *Gata6* RNA within the heart, suggesting translational regulation (Brewer et al., 2002a; Brewer et al., 2002b). During differentiation of animal cap explants containing NCCs, Myocardin-dependent expression of smooth muscle genes acts synergistically with SRF but is antagonized by Gata6 (Barillot et al., 2008). Conditional inactivation of Gata6 in VSMCs or NCCs results in perinatal mortality with a spectrum of OFT defects, demonstrating a cell-autonomous requirement for Gata6 in NC-derived VSMCs. These OFT defects are due to severe reduction of Sema3c levels, suggesting the primary function of Gata6 during cardiovascular development may be to regulate the patterning of the cardiac OFT and aortic arch (Lepore et al., 2006).

14.4 Krox20

Krox20, a C2H2 zinc finger transcription factor related to *Drosophila* Krüppel, has two main functions during vertebrate embryonic development. The first is patterning of the hindbrain and associated NCCs by establishing segmented r3 and r5 domains, and the second is control of Schwann cell (SC) development.

14.4.1 Krox20 in Schwann Cells

Krox20 is expressed in early NCCs, in glial components of the cranial and spinal ganglia, and in NC-derived boundary cap cells (Wilkinson et al., 1989). Developmental transitions during SC development are regulated by Krox20, Oct6, Pax3, Sox10, Creb, and some bHLH factors (Bhatheja and Field, 2006; Jessen and Mirsky, 1998; Jessen and Mirsky, 2002; Kuhlbrodt et al., 1998a; Svaren and Meijer, 2008). Krox20 is not needed during the initial activation of *Myelin* gene expression in developing SCs but is required, together with Oct6, for the transition of SCs from the promyelinating to myelinating stage (Chelyshev Iu and Saitkulov, 2000; Kamholz et al., 1999). Mutations in human Krox20 are associated with Charcot–Marie–Tooth neuropathy type 1 (CMT1), a peripheral nerve demyelinating disease (Kamholz et al., 1999). Krox20 and Pax3 functions are modulated by Sox10 in developing and mature glia (Kuhlbrodt et al., 1998a). Sumoylation of Sox10 modulates its binding synergy with Krox20 and Pax3 on the *Gjb1* and *Mitf* promoters (Girard and Goossens, 2006). SC precursors migrate along growing axons toward their final location. Krox20 is a major target of promyelinating signals, and Krox20 promotes many phenotypic changes in immature SCs that characterize their transition to myelinating cells. Krox20 interacts functionally with neuregulin and TGFβ, two factors implicated in myelination in postnatal ganglia (Jessen and Mirsky, 2002). Neuregulin addition to Schwann cell precursors initiates an increase in cytoplasmic Ca^{2+} that activates calcineurin and the downstream transcription factors Nfatc3 and Nfatc4. Sox10 is an Nfat nuclear partner and synergizes with Nfatc4 to activate Krox20, which regulates genes necessary for myelination (Kao et al., 2009). Jun is required for SC proliferation and death and is downregulated by Krox20 during myelination. Forced expression of Jun in SCs prevents myelination, and in injured nerves, Jun is required for appropriate dedifferentiation, reemergence of immature SC state, and nerve regeneration (Mirsky et al., 2008).

A lacZ knock-in allele of Krox20, initially restricted to boundary cap cells in trunk regions, can be used to trace the fate of these cells and their progeny during development. Trunk boundary cap-derived cells migrate along peripheral axons and colonize spinal root ganglia and DRGs, giving rise to all SC precursors, some neurons (mostly nociceptive), and satellite cells. Boundary cap cells are a source of PNS progenitors that arrive in a secondary wave of migration after the major ventrolateral migratory stream of NCCs (Maro et al., 2004). In vitro, gliogenesis of cultured boundary cap-derived NCSCs (bcNCSCs), monitored by expression of Krox20, Sox2, Sox19, S100, GFAP, and fibronectin, differentiates similarly to that in vivo, sequentially adopting SC precursor and immature SC fates before maturing into myelinating and nonmyelinating SCs (Aquino et al., 2006).

14.4.2 Krox20 in Patterning

Krox20 is involved in segmentation and patterning of part of the hindbrain and its associated NCCs (Nardelli et al., 2003). Krox20 expression is conserved among mice, frogs, fish, and chicken in r3 and r5 and in a stream of NCCs migrating from r5 toward the third branchial arch (Bradley et al., 1993; Ghislain et al., 2003; Nardelli et al., 2003; Nissen et al., 2003; Nonchev et al., 1996; Wilkinson et al., 1989). Expression of Krox20 is downregulated first in r3 and then in r5 (Oxtoby and Jowett, 1993). Krox20 is often used as a marker of the segmented hindbrain, particularly for odd-numbered rhombomeres (Deflorian et al., 2004; Eroglu et al., 2006; Ishii et al., 2005; Kitaguchi et al., 2001; Labosky and Kaestner, 1998; Menegola et al., 2004; Menegola et al., 2005; Nardelli et al., 2003). Expression occurs in precursors in both r5 and r6, and both rhombomeres contribute to Krox20-expressing NC, emigration occurring first from r6 and later caudally from r5. In Krox20 mutants, hindbrain segmentation is disrupted, but at least some prospective r3 and r5 cells persist (Voiculescu et al., 2001). The remaining r3 cells acquire an r2 or r4 identity, and r5 cells acquire an r6 identity. Embryonic chimeras between *Krox20* homozygous-null and wild-type cells show that Krox20 is required for formation of alternating rhombomere identities by coupling segment formation, cell segregation, and specification of regional identity (Voiculescu et al., 2001). In the chicken embryo, transplantation of r6 to the level of r4 or r5 causes many Krox20-expressing cells to migrate rostral to the otic vesicle, but transplantation of r5 to the r4 position results in only a few migrating cells that express Krox20. This demonstrates that Krox20 expression in pharyngeal arch NC does not necessarily correlate with rhombomeric segmentation, and there may be intrinsic differences between the r5 and r6 Krox-20-expressing populations (Nieto et al., 1995).

Krox20 controls the segment-restricted upregulation of Hoxb2 and Hoxa2 (Nonchev et al., 1996; Wilkinson, 1993). *Krox20*-null embryos have a clear loss of Hoxa2 expression in r3, placing

Hoxa2 downstream of Krox20. An enhancer in the Hoxa2 locus with two Krox20 sites is needed for r3/r5 activity, but not NC and mesodermal activity. Ectopic expression of Krox20 in r4 can transactivate the Hoxa2-lacZ reporter (Nonchev et al., 1996). Comparative genomic analysis of the striped bass Hoxb2a-b3a intergenic region to those from zebrafish, pufferfish, human, and mouse demonstrated the presence of a conserved Krox20 binding site-containing enhancer necessary for r3, r4, and r5 expression (Scemama et al., 2002). In addition to regulating Hoxa2 and Hoxb2, Krox20 regulates its own transcription as part of an autoregulatory loop required in expansion and maintenance of Krox20-expressing territories (Voiculescu et al., 2001) A conserved NC enhancer containing two conserved Krox20 binding sites can recapitulate Krox20 NC expression in transgenic mice. This enhancer includes essential Sox binding sites and is strongly activated by Sox10 in synergy with Krox20 to maintain Krox20 expression in the migrating NC (Ghislain et al., 2003). Interactions at three conserved long-range enhancer elements (A, B, and C) also control Krox20 expression. Element A contains Krox20-binding sites, which are required, along with Krox20 protein, for activity. B and C are activated at the earliest stage of Krox20 expression in r5 and r3-r5, respectively, independently of Krox20 binding, initiating Krox20 expression. HNF1b is a direct initiator of Krox20 expression at element B (Chomette et al., 2006).

RA signaling greatly influences Krox20-mediated hindbrain and NCC patterning. Segment-restricted expression of Krox20 responds rapidly to RA treatment and undergoes a progressive series of changes in segmental expression (Marshall et al., 1992). Mouse embryos treated with RA just before differentiation of the cranial neural plate and the start of segmentation have reduction in preotic hindbrain length, loss of rhombomeric segmentation, and NC defects. These defects correlate with changes in Krox20 and Hoxb1 distribution (the r3 domain of Krox20 is absent and the Hoxb1/Krox-20 boundary is poorly defined) (Morriss-Kay et al., 1991). Inactivation of both RAR and RARβ causes expansion of the Krox20 domain into r6 and r7, and defects in structures partially derived from post-otic rhombomere-associated mesenchymal NCCs like the thymus (Dupe et al., 1999). *Raldh2*-null embryos do not have r3- and r5-restricted Krox20 expression but instead a single broad domain of Krox20 (Niederreither et al., 2000).

Expression of Krox20 is modulated by a complex interplay between RA, Wnts, FGFs, and BMPs (Nardelli et al., 2003). In *Xenopus* animal explants, Wnt1 and Wnt3a, expressed in the dorsal NT, synergize with the neural inducers Noggin and Chordin to generate the NC, marked by Krox20, AP-2α, and Slug. Overexpression of Wnt1 or Wnt3a in the neuroectoderm of whole embryos led to a dramatic increase of Slug and Krox20-expressing cells, but hindbrain expression of Krox20 remained unaffected (Saint-Jeannet et al., 1997). Adding FGFs to the chicken NT causes ectopic Krox20 expression in caudal hindbrain NC (r7 and r8) and expansion of the Krox20-expressing area in the neuroepithelium. Application of an FGF pathway inhibitor leads to downregulation of Krox20 in the hindbrain neuroepithelium and NC (Marin and Charnay, 2000).

14.5 Mecom/Evi1

Mecom/Evi1 encodes a putative protooncogenic transcription factor containing 10 zinc finger motifs and is associated with myeloid neoplasia, myeloid leukemia, and myelodysplasia (Hoyt et al., 1997; Mead et al., 2005). Mecom is expressed in embryonic mesoderm, pronephric tissue, primary head folds, and NC-derived cells associated with the PNS (Hoyt et al., 1997; Kazama et al., 1999; Mead et al., 2005). Mice lacking full-length Mecom die at approximately E10.5 with widespread hypocellularity, hemorrhaging, and disrupted paraxial mesenchyme development, defects in the heart, somites, cranial ganglia, and peripheral nervous system (Hoyt et al., 1997). Mecom may also be important for general regulation of early neuroectodermal differentiation (Kazama et al., 1999).

14.6 Nczf

Nczf (neural crest zinc finger), a target gene of Tlx2, encodes a novel Krüppel-associated box zinc finger transcriptional repressor. The Nczf consensus binding sequence is found within regulatory regions of *Ednrb* and *Mitf* loci, suggesting a possible role for Nczf as a sequence-specific regulator of NC development (Kitahashi et al., 2007).

14.7 Prdm1 AND Prdm2

The PR domain transcriptional repressors Prdm1 and Prdm2 contain SET/zinc-finger domains. Murine Prdm1, also known as B lymphocyte-induced maturation factor (Blimp1), functions as a master regulator of B-cell terminal differentiation. *Prdm1*-null embryos die at mid-gestation with placental and pharyngeal arch defects, but despite its role in NC induction, NC formation is unaffected (Vincent et al., 2005). Zebrafish Prdm1 is induced by BMP signaling in cells at the boundary of the neural plate and nonneural ectoderm (Hernandez-Lagunas et al., 2005; Roy and Ng, 2004). Loss of Prdm1 activity inhibits specification of the NC and primary sensory neuron development, resulting in a reduction in NCCs and complete loss of Rohon-Beard sensory neurons. Overexpression of Prdm1 in mutants rescues both phenotypes, and in wild-type zebrafish, results in the generation of supernumerary default-fate primary sensory neurons. BMP-induced Prdm1 activity is needed for cell fate specification of both NCCs and primary sensory neurons (Hernandez-Lagunas et al., 2005; Roy and Ng, 2004). Pheochromocytomas and abdominal paragangliomas are rare tumors from NC-derived chromaffin cells. The *Prdm2* tumor suppressor locus is in a region of chromosome 1 frequently deleted in these tumors. Loss of heterozygosity at the *Prdm2* locus was detected in most of the tumors, and Prdm2 mRNA was downregulated in tumors compared to normal adrenal control (Geli et al., 2005).

14.8 SNAIL

Vertebrate Snail is a zinc finger transcription factor expressed in NC and mesoderm and orthologous to *Drosophila* Snail. It regulates EMT during gastrulation, and this function is at least partially conserved in EMT of the NC. Early in NC development, the two main functions of Snail are to promote delamination during EMT and to suppress apoptosis (Figure 14.1).

14.8.1 *Xenopus* Snail

In *Xenopus*, Snail is expressed in the developing neural folds at the neural plate border (NPB) in what will become the NC and roof plate; this expression is maintained in migratory NCCs (Gadson et al., 1993; Linker et al., 2000; Mayor et al., 1995). Snail and its paralog Slug are among the earliest genes that respond to NC-inducing signals from the mesoderm of the dorsal or lateral marginal zone (LaBonne and Bronner-Fraser, 2000; Mayor et al., 1995). Snail expression precedes Slug expression, and Slug expression precedes expression of Twist, another early NC transcription factor (Aybar et al., 2003; Linker et al., 2000). One of the factors upstream of Snail is Msx1. Expression of dominant-negative Msx inhibits Snail and other early NC markers, and resulting NC defects can be rescued by injection of Snail or Slug (Tribulo et al., 2003). Conditional Snail gain-of-function and loss-of-function demonstrate that Snail is upstream of Slug and induces expression of Slug and other NC markers (Zic5, Foxd3, Twist and Ets1), indicating a key role for Snail in NC specification and migration (Aybar et al., 2003). At the NPB, Snail represses and therefore restricts the expression of Delta1, which interacts with Notch to activate the Bmp4 repressor Hes4a; modulation of Bmp4 levels by Snail provides permissive conditions for specification of NPB cells (Glavic et al., 2004b). Snail protein generally acts as a transcriptional repressor and complexes with Jub (an Ajuba Lim protein) through a SNAG repression domain to silence transcription of genes such as Cdh1 (E-cadherin), a molecule important for NC delamination (Langer et al., 2008).

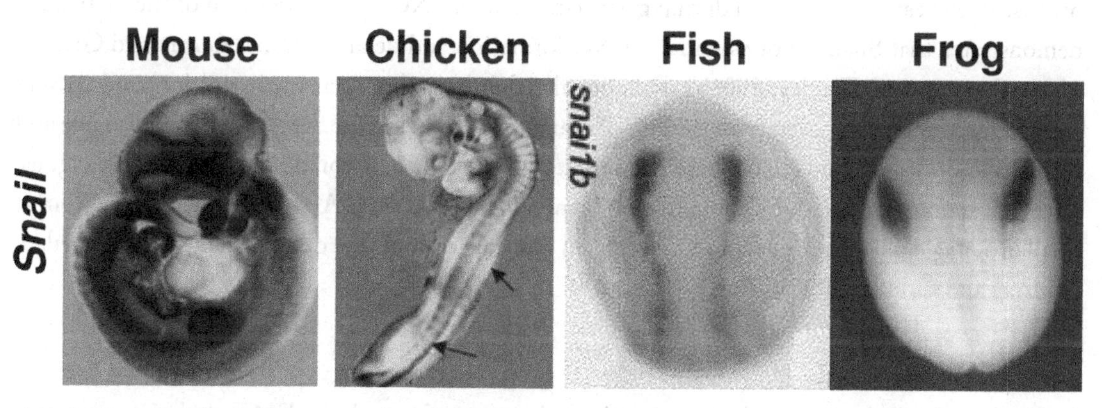

FIGURE 14.1: Whole-mount in situ hybridizations show *Snail* in mouse, chicken, fish, and frog embryos. Mouse: Sefton et al., 1998, 10.5-dpc embryo, lateral view; chicken: Sefton et al., 1998, stage 18 embryo, lateral view; zebrafish: Rau et al., 2006, three-somite embryo, dorsal view; *Xenopus*: Schuff et al., 2007, stage 23 embryo, dorsal view.

14.8.2 Zebrafish Snail

In zebrafish, Snail is expressed in the entire NC and in mesodermal derivatives of the head region (Hammerschmidt and Nusslein-Volhard, 1993). Although Snail is an early NC marker, it is downstream of several other, earlier NC transcription factors such as Sox9 and Foxd3 (Montero-Balaguer et al., 2006; Stewart et al., 2006; Yan et al., 2005). Two zebrafish Foxd3 mutants, $\mathit{Foxd3}^{sym1}$ and $\mathit{Foxd3}^{mother\ superior}$, when homozygous, show reduced levels of Snail expression in premigratory NCCs and subsequent depletion of many NC derivatives (Montero-Balaguer et al., 2006; Stewart et al., 2006).

14.8.3 Chicken Snail

During early stages of development, many expression domains of Slug and Snail in mouse compared to chicken embryos are reversed, but in later NC development, the sites of expression of Slug and Snail are conserved between these two species (Sefton et al., 1998). In early NC development of the chicken embryo, Slug performs the main functions that Snail carries out in the mouse, and chicken Snail is not expressed in the premigratory NC. However, electroporation experiments in the chicken embryo have shown functional equivalence between both Slug and Snail genes of both species (del Barrio and Nieto, 2002).

14.8.4 Mouse Snail

Snail is expressed in the mouse premigratory NC. Mouse Snail, together with Sox9, which provides competence for NCCs to undergo EMT, is sufficient to induce an EMT in neural epithelial cells (Cheung et al., 2005). In mouse embryos, a homozygous null mutation in Snail is embryonic lethal because of the early role of Snail during gastrulation, but an NC-specific deletion of the *Snail* gene demonstrates that Snail is not essential for NC formation and delamination (Murray and Gridley, 2006a; Murray and Gridley, 2006b). Functional redundancy between Snail and its paralog Slug seems to account for at least part of the lack of dramatic NC phenotype. *Slug*-null mice are viable and fertile, and only a percentage of these exhibit cleft palate, whereas compound *Snail*-heterozygous, *Slug*-null mutant mice exhibit a completely penetrant cleft palate. Also, NC-specific deletion of Snail in a *Slug*-null genetic background results in multiple craniofacial defects, including cleft palate (Murray and Gridley, 2006a; Murray et al., 2007).

14.8.5 Human Snail

In humans, Snail function is highly significant because of its role in EMT during tumor progression (Okajima et al., 2001; Paznekas et al., 1999; Twigg and Wilkie, 1999). Studies of Snail in tumorigenesis of multiple tissue types, including those derived from the NC, reveal that Snail is expressed in these tumors and that Snail-induced EMT is associated with direct repression of Cdh2(E-cadherin) transcription (Blanco et al., 2002; Fendrich et al., 2009; Kuphal et al., 2005; Locascio et al., 2002). Snail expression may affect cellular differentiation (Takahashi et al., 2004), and levels of expression inversely correlate with the grade of tumor differentiation (Blanco et al., 2002). The progression of NC-derived cancers, such as melanoma, has been linked to expression of Snail (Fendrich et al., 2009; Kuphal et al., 2005; Palmer et al., 2009). Knockdown of Snail in a melanoma cell line results in downregulation of the EMT-associated genes Mmp2, Emmprin, Sparc, Timp1, Tpa, RhoA, and Notch4, and reinduction of Cdh1 (Kuphal et al., 2005). Confirmed binding sites for Ets1/Yy1 and Sox recognition motifs are present in a conserved 3' enhancer of the *Snail* gene, and both of these are needed for full activity in melanoma cells. Yy1 is needed for Snail expression; it is present only in cells expressing Snail, and Yy1 knockdown leads to downregulation of Snail (Palmer et al., 2009).

14.9 SLUG

Slug, also known as Snail2, is highly similar to Snail and in many instances, as described below, can substitute for Snail because of its functional redundancy. The function of Slug seems to vary greatly in vertebrates; in mouse, Slug is not expressed in the premigratory NC, and NC specification functions are carried out by Snail alone. In the chicken embryo, Slug is expressed in premigratory NC and seems to cover many of the functions of Snail (Figure 14.2).

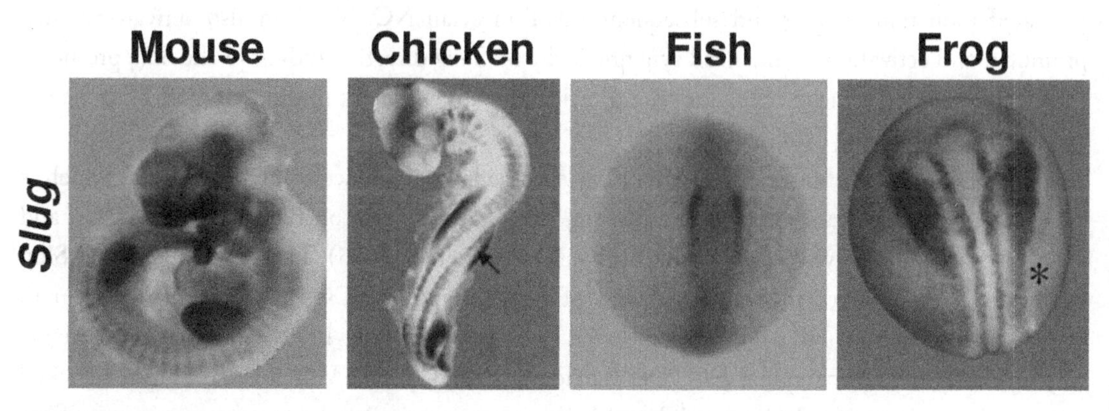

FIGURE 14.2: Whole-mount in situ hybridizations show *Slug* in mouse, chicken, fish, and frog embryos. Mouse: Sefton et al., 1998, 10.5-dpc embryo, lateral view; chicken: Sefton et al., 1998, stage 18 embryo, lateral view; zebrafish: Eroglu et al., 2006, 12-hpf embryo, dorsal view; *Xenopus*: O'Donnell et al., 2006, stage 13/14 embryo, dorsal view.

14.9.1 Avian Slug

In the chicken embryo, Slug is expressed in the NC, and knockdown of Slug specifically inhibits NC EMT (Nieto et al., 1994). Transplantation experiments show that the expression of Slug in addition to other dorsal markers Wnt1 and Wnt3a, and subsequent formation of NCCs, can be induced by the juxtaposition of nonneural and neural ectoderm (Dickinson et al., 1995). When the avian neural folds are ablated, the NC population can be replaced by the remaining neuroepithelial cells. Slug expression is an early response in the reformation of the NC and begins shortly after midline closure and before emergence of NCCs (Buxton et al., 1997; Sechrist et al., 1995). During the course of NC formation, Slug expression, as a marker for NC, can be repressed by engraftment of notochords or Shh-secreting cells, suggesting that prospective NCCs in the open neural plate are not yet committed to an NC fate (Selleck et al., 1998). Neural folds derived from all rostrocaudal levels of the open neural plate are competent to express Slug, but in the absence of tissue interactions, Slug becomes downregulated, indicating that additional signals are needed for maintenance of Slug expression (Basch et al., 2000).

Noggin and Bmp4 in the dorsal NT may trigger delamination of specified, Slug-expressing NCCs (Sela-Donenfeld and Kalcheim, 1999). During NC formation, quail Deltex2 is expressed throughout the ectoderm. Misexpression of a dominant-negative form of Deltex in the ectoderm inhibited expression of the NC marker Slug and NC inducer Bmp4, and cotransfection of Bmp4 can rescue Slug expression (Endo et al., 2003). In the anterior ectoderm, BMP signaling is needed for Slug expression (Sakai et al., 2005). Slug and subsequent EMT are effectively induced by BMP4, and phosphorylated Smad1 is detected in the neural plate and the neural folds (Sakai et al., 2005). Gain-of-function and loss-of-function experiments reveal that Sox9 is essential for BMP signal-mediated induction of Slug and subsequent EMT in avian NC. Slug can also activate its own promoter, and activation is enhanced synergistically by Sox9 directly binding to the *Slug* promoter (Sakai et al., 2006). Chicken Slug, in synergy with Sox9, is sufficient to induce an EMT in neural epithelial cells (Cheung et al., 2005).

A major role for Slug in the early NC is specifying competence to undergo EMT, which is associated with a decrease in cell–cell adhesion, loss of N-cadherin on the surface of NCCs, and changes in integrin activity as migration begins (Duband et al., 1995). TUNEL labeling and Slug expression were observed in a subpopulation of early migrating NCCs and presumably occur in subpopulations of both NC and neuroepithelial cells (Lawson et al., 1999). Slug marks premigratory and early migratory cranial NC (Del Barrio and Nieto, 2004). When avian Slug is knocked down in the premigratory NC, several dorsal NT genes are upregulated due to derepression of Slug targets. Cdh6, one of the target genes, contains three clustered E-box pairs that Slug binds directly to regulate Cdh6 transcription (Taneyhill et al., 2007). Cdh6 is expressed in the dorsal NT before NC emigration but is then repressed by Snail2 in premigratory and early migrating cranial NCCs.

Knockdown of Cdh6 leads to premature NCC emigration, and Cdh6 overexpression disrupts migration (Coles et al., 2007). Slug cooperates with Ets1, specifically expressed in cranial NCCs, which is responsible for mobilization of premigratory NCCs (Theveneau et al., 2007).

14.9.2 *Xenopus* Slug and Upstream Regulation

A large body of research in *Xenopus* has helped to reveal the complex regulatory network involved in NC induction and specification, identifying many factors upstream of Slug. In *Xenopus*, Slug, along with its ortholog Snail, is one of the earliest markers of the developing NC (LaBonne and Bronner-Fraser, 1998; LaBonne and Bronner-Fraser, 2000). Slug expression, along with the transcription factors AP-2 and Krox20, marks the prospective NC at the dorsal most part of the NT, and expression of these genes seems to be controlled by Wnt1, Wnt3a, noggin, and chordin (Saint-Jeannet et al., 1997). However, Slug overexpression is not sufficient to induce NC. In addition to Slug, inhibition of BMP signaling seems necessary, but this requirement can be bypassed in the presence of Wnt or FGF signaling, allowing Slug-expressing ectoderm to form NC (LaBonne and Bronner-Fraser, 1998). During NC specification, Slug expression can also be induced by Snail. Slug alone is unable to induce other NC markers in animal cap assays, and Snail and Slug are functionally equivalent when tested in overexpression studies, suggesting that Snail can carry out some functions previously associated with Slug. This is supported by rescue experiments in embryos injected with dominant-negative constructs, indicating that Snail lies upstream of Slug in the genetic cascade leading to NC formation and that it plays a key role in crest development (Aybar et al., 2003). Slug, Snail, and Twist all are expressed in the presumptive neural folds and mark the presumptive NC and roof plate, initiating expression sequentially, with Snail preceding Slug, which precedes Twist. At the beginning of gastrulation, Snail is present in a unique domain of expression in a lateral region of the embryo in both superficial and deep layers of the ectoderm, as are Slug and Twist (Linker et al., 2000). Slug is expressed specifically in the prospective NC region and continues in pre- and postmigratory cranial and trunk NC. Slug, like Snail, is also induced by dorsal or lateral marginal zone mesoderm. Explants of the prospective NC are competent to express Slug (Mayor et al., 1995). Slug is specified later than Snail, which is not detected until stage 12 (Mayor et al., 1995). Wnt signaling is implicated in direct Slug regulation during NC formation because two regions of the *Xenopus* Slug promoter, one of which contains an essential Lef/B-catenin binding site, are necessary and sufficient to drive NCC-specific expression (Vallin et al., 2001). Downstream of signaling pathways, an NC regulatory network consisting of Sox10, Foxd3, and Slug seems to be one of the driving forces that shape the early NC (Pohl and Knochel, 2001).

Slug marks the prospective NC (Bastidas et al., 2004). Msx1 and Msx2 are required for NC Slug expression (Khadka et al., 2006; Tribulo et al., 2003). In the developing neural tube, Msx1 induces Pax3 and Zic1 cell autonomously, then Pax3 combined with Zic1 activates Slug in a Wnt-dependent manner (Monsoro-Burq et al., 2005; Sato et al., 2005). Loss of either Msx1 or Msx2 leads to changes in neural and epidermal expression boundaries, but loss of one gene can be compensated for by overexpression of the other (Khadka et al., 2006).

14.9.3 Downstream of *Xenopus* Slug

As described for Snail, the Slug protein also contains a conserved SNAG repression domain required for the assembly of a repressor complex and interacts directly with Ajuba LIM corepressor proteins to silence targets such as Cdh2 (E-cadherin) (Langer et al., 2008). In the developing neural folds, apoptosis is more prevalent than in the rest of the neural ectoderm, and Slug, as an antiapoptotic factor, along with Msx1, which promotes cell death, have opposing effects, such that there is less apoptosis in the prospective NC than in the region of the neural folds adjacent to the NC, where Msx1 expression is maintained. Overexpressing Slug can protect cells from apoptosis, and this effect can be reversed by expressing the apoptotic factor Bax. Slug and Msx1 control transcription of Bcl2 and several Caspases required for apoptosis (Tribulo et al., 2004), and an absence of NC marker expression after Slug knockdown can be rescued by expression of the antiapoptotic protein Bcl-xL. Bcl-xL's effects are dependent on IκB kinase-mediated activation of the bipartite transcription factor NF-κB. NF-κB, in turn, directly upregulates *Slug* and *Snail* transcription. Slug indirectly upregulates NF-κB subunits and directly downregulates transcription of proapoptotic *Caspase9* (Zhang et al., 2006a). This antiapoptotic function of Slug allows it to control cell numbers among developing NC derivatives (Tribulo et al., 2004).

Experiments fusing the Slug DNA binding domain with activation or repression domains demonstrate that Slug acts as a repressor during NC generation (LaBonne and Bronner-Fraser, 2000; Mayor et al., 2000). Overexpression of Slug leads to expanded expression of NC markers and an excess of melanocytes (LaBonne and Bronner-Fraser, 2000). Slug antisense RNA did not suppress early expression of the related gene Snail but led to reduced expression of both Slug and Snail in later-stage embryos, whereas the expression of another NC marker, Twist, was not affected. Downregulation of Slug and Snail coincided with inhibition of NCC migration and reduction or loss of many NC derivatives. In particular, formation of rostral cartilages was often highly aberrant, whereas the posterior cartilages were less frequently affected (Carl et al., 1999). The effects of Slug antisense RNA on NC migration and cartilage formation defects could be rescued by the injection of Slug or Snail mRNA. Slug is required for NC migration and Snail may be functionally redundant, and both genes are required to maintain each other's expression in NC development (Carl et al., 1999).

Slug overexpression in the NT of the chicken embryo increases Rhob expression, the number of HNK-1-positive migratory cells, and cranial NC production (del Barrio and Nieto, 2002). Rhov is an early expressed NC marker essential for NCC induction. RhoV mRNA is maternally expressed and accumulates shortly after gastrulation in the NC forming region. RhoV depletion impairs expression of the NC markers Sox9, Slug, or Twist but has no effect on Snail induction (Guemar et al., 2007). Ectopic expression of both dominant-negative and constitutively active Rho GTPase mutants after RNA or DNA microinjection disrupted the endogenous expression of Slug, Foxd3, and Snail. Constitutively active RhoA was inhibitory, whereas dominant-negative RhoA increased NC marker gene expression. Rho GTPases can regulate the expression of Slug during NC ontogeny (Broders-Bondon et al., 2007).

14.9.4 Mouse Slug

When the mouse *Slug* gene was identified, cloned, and characterized (Jiang et al., 1998a; Jiang et al., 1998b; Rhim et al., 1997; Savagner et al., 1998; Sefton et al., 1998), its expression pattern was strikingly different from that of other studied vertebrates. Unlike in avians and amphibians, mouse Slug is not expressed in premigratory NCCs but rather only in migratory NCCs (Jiang et al., 1998a; Savagner et al., 1998). Many of the sites of Slug and Snail are reversed when comparing rodents to birds and amphibians; however, the later NC expression of both Slug and Snail seems to be more conserved among all vertebrates (Sefton et al., 1998). In both mice and rats, the earliest significant expression of Slug is in cranial NCCs that are migrating and invading the first pharyngeal arch, and Slug expression is maintained in craniofacial NC derivatives (Savagner et al., 1998). Surprisingly, *Slug*-null embryos have normal NC formation, migration, and differentiation, demonstrating that neither expression pattern nor biological function of Slug are conserved (Jiang et al., 1998a; Murray and Gridley, 2006b). *Slug* homozygous-null mice are mostly viable and fertile (Murray and Gridley, 2006a), with only melanocyte defects, making the *Slug* homozygous-null mouse a model for type 2 Waardenburg syndrome (WS2) (Perez-Mancera et al., 2006; Tachibana et al., 2003), and a few mice also have cleft palate (Murray et al., 2007). Compound *Snail*-heterozygous, *Snail*-null embryos have a completely penetrant cleft palate, and NC-specific *Snail* deletion in a *Slug*-null background also results in multiple craniofacial defects, including cleft palate (Murray et al., 2007). Analysis of a lacZ knock-in into the *Slug* locus shows that Slug is highly expressed in the craniofacial mesenchyme and NC-derived craniofacial skeleton as well as in the OFT and endocardial cushions of the heart (Oram et al., 2003). Slug is critical for the development of many NC-derived cells. Overexpression of Slug in mice carrying a Slug transgene does not cause any apparent developmental defects, but these adult mice have a 20% incidence of sudden death as well as cardiac failure that may be associated with progression and proliferation of mesenchymal tumors (Perez-Mancera et al., 2006; Perez-Mancera et al., 2005).

In zebrafish, Slug mRNA accumulates at the neural plate border including the prospective NC, and expression later becomes restricted to the NC. Slug is one of the earliest NC-specific genes known in zebrafish (Thisse et al., 1995).

14.9.5 Slug in Humans

Patients with WS2 have been identified with homozygous deletions in Slug leading to absence of Slug protein, demonstrating the requirement of Slug in the melanocyte lineage (Hou and Pavan, 2008; Sanchez-Martin et al., 2002; Sanchez-Martin et al., 2003; Tachibana et al., 2003). This role is supported by a number of cases of human piebaldism without Kit mutations that have been attributed to Slug mutations (Tachibana et al., 2003). Mitf can transactivate the Slug promoter, and a genetic interaction exists between Slug and Kit (Sanchez-Martin et al., 2002). Slug and several other NC regulatory genes can be found in primary human melanocytes that form melanomas and is required for metastasis of transformed melanoma cells (Gupta et al., 2005). Slug duplication is associated with tetralogy of Fallot and cleft palate among other defects (Perez-Mancera et al., 2006).

14.9.6 Snail and Slug in Stem Cell Culture

During the directed differentiation of rat cortical neuron stem cells into a variety of NC lineages, Slug is one of the induced premigratory NC markers (Gajavelli et al., 2004). Multipotent stem cells derived from rat periodontal ligament and porcine skin-derived precursors (SKPs) expressed Slug as well as other NC markers (Techawattanawisal et al., 2007; Zhao et al., 2009). Snail is one of several NC markers induced in the formation of NC-like cells during the differentiation of human embryonic stem cells (hESCs) toward a neuronal lineage in the presence of mouse stromal cells (Pomp et al., 2005). Slug and Snail are both among the NC markers that are expressed in mouse dental NC-derived progenitor cells (Degistirici et al., 2008).

14.10 Zeb1 AND Zeb2

The zinc finger E-box binding homeobox (Zeb) transcriptional repressor family has two factors expressed in NC: Zeb1 (also known as deltaEF1) and Zeb2 (also known as Smad-interacting protein 1/SIP1). Zeb1 is a transcriptional corepressor of Smad target genes with functions in the patterning of NC-derived cells, CNS, and midline structures. Mutations in Zeb1 can lead to neurological disorders in addition to dysmorphic features, megacolon, and other malformations (Sztriha et al., 2003). A human Zeb1 mutation linked to Hirschsprung's disease suggests that deficiency of Zeb1, resulting in disrupted BMP/Smad signaling, may contribute to ENS defects (Cacheux et al., 2001). Zeb2 is expressed in NC derivatives, notochord, somites, limbs, and a few domains of the CNS. *Zeb2*-null mice die perinatally with severe T-cell deficiency of the thymus and various skeletal defects, including defects in the NC-derived craniofacial skeleton (cleft palate, Meckel's cartilage hyperplasia, nasal septum dysplasia, shortened mandible) (Takagi et al., 1998). *Xenopus* Zeb1 is initially strongly expressed in prospective neurectoderm and later continues primarily within the NT and NC (van Grunsven et al., 2000). In *Xenopus*, Zeb1 and Zeb2 are coexpressed in migratory cranial NC, in the retina, and in the NT. Overexpression of Zeb2, like Zeb1, reduces expression of BMP-dependent genes, but only Zeb1 induces neural markers. Zeb1 and Zeb2 both form complexes with Smad3, coactivators p300 and pCAF, and the corepressor Ctbp1, but have different repression efficiencies (van Grunsven et al., 2006).

14.11 Zfp704

Zinc finger protein 704, also known as glucocorticoid-induced gene-1 (Gig1), is a C2H2 zinc finger protein. In mouse embryos, X-gal staining of a *Zfp704*$^{nuclear\ lacZ}$ knock-in shows expression in several NC-derived lineages (Yamamoto et al., 2007).

14.12 Zic1

The *Zic* genes are vertebrate orthologs of the *Drosophila* pair rule gene *odd-paired* encoding C2H2 zinc finger transcription factors (Kuo et al., 1998). There are five Zic orthologs (Zic1–Zic5) conserved in mammals, birds, frogs, and fish (Inoue et al., 2007; Keller and Chitnis, 2007; Nakata et al., 1998; Warner et al., 2003), and zebrafish has a Zic6 gene lacking a subset of expression domains typical of other Zic family members (Keller and Chitnis, 2007). The *Zic* genes are generally expressed in the dorsal NT and play multiple roles during neural development: NC induction, neural differentiation, inhibition of neurogenesis, left–right asymmetry, and regulating proliferation (Elms et al., 2003; Keller and Chitnis, 2007; Merzdorf, 2007; Warner et al., 2003). Although the *Zic* genes have broadly overlapping expression, mutants for individual *Zic* genes typically have distinct phenotypes (Inoue et al., 2007). *Zic1*, *Zic2*, and *Zic3* are very similar to each other (Nakata et al., 1998), whereas *Zic4* is very close to *Zic5* (Fujimi et al., 2006). Zic proteins can interact physically with other Zic proteins or regulate Zic transcription, indicating a large potential for crosstalk among Zic family members (Fujimi et al., 2006; Keller and Chitnis, 2007).

Zic1 encodes a transcriptional activator expressed throughout the presumptive neural plate that becomes restricted to the dorsal NT and NC (Kuo et al., 1998; Nakata et al., 1998). Zic1 plays a role in patterning the early neural plate and in formation of the NC, somites, and cerebellum (Cornish et al., 2009; Kuo et al., 1998). Zic1 is also one of the earliest molecular indicators of neural fate determination in *Xenopus* (Tropepe et al., 2006); it renders animal cap ectoderm competent to respond to the neural inducer Noggin and modifies the array of genes induced by Noggin. Activated Zic1 induces NC and dorsal NT markers in both animal caps and whole embryos. In more ventral ectoderm, Zic1 induces formation of loose cell aggregates suggestive of NC precursor cells as it suppresses ventral fates (Kuo et al., 1998).

In neuralized ectoderm, Msx1 induces Pax3 and Zic1, and then Pax3 and Zic1 activate Slug in a Wnt- and Foxd3-dependent manner (Monsoro-Burq et al., 2005; Sasai et al., 2001; Sato et al., 2005). Pax3 and Zic1 expression overlaps in the presumptive NC area before NC marker genes

Foxd3 and Slug, and this coexpression is essential for initiating NC differentiation (Sato et al., 2005). Overexpression of Zic1 and Pax3 together induces ectopic NC differentiation in ventral ectoderm, but overexpression of either alone expands NC marker expression only within the dorsolateral ectoderm (Nakata et al., 1998; Sato et al., 2005). In *Xenopus*, Pax3 and Zic1 are necessary and sufficient to promote hatching gland and preplacodal fates, respectively, whereas their combined activity is essential to specify the NC (Hong and Saint-Jeannet, 2007). Inhibition of BMP signaling is critical for activation of *Xenopus* Zic1 expression and establishing neural identity (Tropepe et al., 2006). Zic1 is not expressed in migratory NC (Sun Rhodes and Merzdorf, 2006). In the chicken embryo, Zic1, Zic2, and Zic3 are expressed in the hindbrain, NC, periotic mesenchyme, and inner ear. Zic1 and Zic2 have very similar expression patterns and strength (Zic2 is generally slightly weaker), and Zic3 has very weak expression (Warner et al., 2003).

14.13 Zic2

Murine Zic2 is first expressed in the entire neural plate, becomes restricted to the lateral region, and is required for timing of neurulation. Knockdown of murine Zic2 delays neurulation, resulting in holoprosencephaly (HPE, also associated with human Zic2 mutations) (Houston and Wylie, 2005; Nagai et al., 2000), spina bifida, perinatal mortality, delayed differentiation of the dorsal-most neural plate (roof plate and NCCs), and impaired development of NC derivatives such as the DRGs (Nagai et al., 2000). A *Zic2* hypomorphic allele or loss-of-function delays NC production resulting in a decrease of NCCs. In addition, Zic2 contributes to hindbrain patterning and is essential for normal development of r3 and r5 (Elms et al., 2003). In *Xenopus*, Zic2 is expressed broadly in the ectoderm (Nakata et al., 1998), then becomes restricted to stripes that alternate with regions giving rise to primary neurons, and inhibits neurogenesis and induces NC differentiation by counteracting the neurogenic inducing activity of Gli proteins (Brewster et al., 1998). RA and Shh signaling have opposite effects on Gli3 and Zic2 that prepattern the neural plate and RA cannot rescue the inhibitory effect of Zic2 on primary neurogenesis. RA treatment inhibits the expression of *Xenopus* Zic2 and Shh in the neural ectoderm while expanding the Gli3 expression domain, but a retinoid antagonist induces Zic2 and Shh. Shh overexpression enlarges the neural plate at the expense of the NC and upregulates Zic2 while downregulating Gli3 (Franco et al., 1999). Overexpression of Zic2 enhanced neural and NC-derived tissue formation in embryos and in animal cap explants (Nakata et al., 1998).

14.14 Zic3

Xenopus Zic3 plays a significant role in neural and NC development (Nakata et al., 1997; Zhu et al., 2007). Zic3 is induced by inhibiting Bmp4 and is first detected in the prospective neural plate following Chordin expression, earlier than most proneural genes. Overexpression of Zic3 results in hyperplastic neural and NC-derived tissue in vivo and induces expression of proneural and NC marker genes in animal cap explants (Nakata et al., 1997). Overexpression of Meis1b in animal cap explants induces Zic3 and can also induce ectopic NC-derived tissues (pigmented cell masses) in developing embryos (Maeda et al., 2001). Zic3, which contains several AP-1 binding sites, is also induced by AP-1 in a dose-dependent manner. Knockdown of the AP-1 component Jun blocks Activin-induced Zic3 expression, whereas coinjection of *Jun* mRNA rescues downregulated Zic3 expression (Lee et al., 2004a). Mutations in *Zic3* cause situs ambiguus or isolated congenital heart defects in humans, such as transposition of the great arteries (Chhin et al., 2007; Zhu et al., 2007). Mutant Zic3 protein is expressed at lower levels, is degraded more quickly, has decreased transcriptional activation of a TK-luciferase reporter, and reduced activity in left–right asymmetry and NC induction in *Xenopus* embryos (Chhin et al., 2007). *Zic3*-null mice also have left–right asymmetry defects, associated cardiovascular, vertebra/rib, and CNS malformations, and 20% lethality but no detectable defects in NC-derived tissues, presumably because of redundancy of other Zic proteins in the NC (Zhu et al., 2007).

14.15 Zic4 AND Zic5

Xenopus Zic4 expression is detected mainly in the NPB, dorsal NT, and somites, similar to Zic1. Injection of *Zic4* mRNA caused the induction of NC marker genes, excess pigment cell formation, and excess neural tissue, suggesting Zic4 can induce neural and NC tissue similar to other Zic family members. The zinc-finger domain is critical for many Zic4 functions, but the C-terminal regions of Zic4 and Zic5 are distinct from Zic1, Zic2, and Zic3 in their involvement in inducing NC genes Slug and Sox10 (Fujimi et al., 2006). Zic5, like the other Zic proteins, is expressed in the early NC and is sometimes used as an NC marker (Aybar et al., 2003).

• • • •

CHAPTER 15

Other Miscellaneous Genes

15.1 Atf1

Clear cell sarcoma of tendons and aponeuroses, also referred to as malignant melanoma of soft parts, is a rare NC-derived cancer. Most cases have a translocation that creates a unique chimeric *Ewsr1/activating transcription factor 1* (*Atf1*) fusion gene transcript (Dim et al., 2007).

15.2 Cited2 AND Cited4

The Crebbp/Ep300-interacting transactivator with Glu(E)/Asp(D)-rich carboxy-terminal domain 2 (Cited2) transcription factor binds Ep300 and its paralog Creb-binding protein (Crebbp), ubiquitously expressed transcriptional coactivators and histone acetyl transferases, with high affinity to regulate transcription. Cited2 is required for NC and NT development (Bamforth et al., 2001; Weninger et al., 2005). *Cited2*-null embryos die with cardiac malformations, adrenal agenesis, abnormal cranial ganglia and exencephaly. Cardiac defects include some typical NC-related defects such as double outlet right ventricle, persistent truncus arteriosus, and right-sided aortic arches (Bamforth et al., 2001). Cited2 interacts with and coactivates all isoforms of AP-2, and *Cited2*-null embryos have increased apoptosis in the midbrain region and a marked reduction in cardiac NCCs expressing the AP-2 target Erbb3 (Bamforth et al., 2001; Weninger et al., 2005). Cited2 is also expressed in the AV canal and cardiac septa, and there are some NC-independent cardiac defects such as ventricular septal defects and hypoplasia of the atrioventricular endocardial cushions evident in *Cited2*-null embryos (Weninger et al., 2005). Heterozygous human *Cited2* mutations are also associated with congenital heart disease (MacDonald et al., 2008).

Human Cited4 and Ep300/Crebbp are present in endogenous interaction complexes. Cited4 functions as a transcriptional activator and physically interacts with all AP-2 isoforms as a coactivator in vitro but is slightly weaker than Cited2 for coactivation of AP-2γ. Cell-type and AP-2 isoform-specific coactivation by different Cited proteins may be a mechanism for differential modulation of AP-2 function (Braganca et al., 2002).

15.3 Creb1

Creb1 (cyclic AMP-responsive element binding protein 1) is a basic leucine zipper transcription factor that mediates cAMP signaling by binding downstream targets at specific conserved cAMP response element (CRE) sites. In NC cultures, cAMP signaling induces Mitf and melanogenesis. A dominant-negative Creb1 inhibits Mitf expression and melanogenesis, supporting the notion that Creb activation is necessary for melanogenesis. However, constitutively active Creb1 alone is not sufficient for Mitf expression and melanogenesis, which requires simultaneous PKA signaling (Ji and Andrisani, 2005). Combined Bmp2 and cAMP signaling induces the catecholaminergic lineage in NC cultures from avian and murine cells by a direct Creb1-mediated increase in expression of *Phox2a*, which has two CRE sites in its promoter (Benjanirut et al., 2006; Chen et al., 2005). In NC cultures, a dominant-negative Creb1 suppresses *Phox2a* transcription and sympathoadrenal lineage development. Constitutively active Creb1 alone can induce *Phox2a* transcription, but it is not sufficient for sympathoadrenal lineage development, which also requires activation of Phox2a by phosphorylation through PKA (Chen et al., 2005). In primary NCSC culture, Gata3 transactivates the *Th* promoter, which does not contain a Gata3 binding site but does have a Creb1 binding site required for Gata3 transactivation. Gata3 physically interacts with Creb1 in vitro and in vivo (Hong et al., 2006). Creb1, along with Oct6, Krox20, Pax3, and Sox10, is also hypothesized to be involved in the developmental transitions of Schwann cells (Jessen and Mirsky, 1998).

15.4 Ctbp1 AND Ctbp2

In *Xenopus* NC, C-terminal binding protein 1 (Ctbp1) forms a complex with either Zeb1 or Zeb2, acting as a corepressor for the targets of these proteins (van Grunsven et al., 2006). Ctbp2, along with Hdac1, is recruited by Sox5, which functions in the melanocyte lineage to modulate Sox10 activity by binding the regulatory regions of melanocytic Sox10 target genes, thus directly competing against Sox10 and recruiting corepressors such as Ctbp2 and Hdac1 (Stolt et al., 2008).

15.5 Dawg

In *Xenopus*, the Dawg (Dachshund-like with gill expression) protein, a member of the Ski/Sno/Dac family containing a putative Ski DNA-binding domain, is expressed in the brain, sensory vesicles, and cranial NC of neurula and tailbud embryos (Seufert et al., 2005). Dawg NC function has not been tested.

15.6 Mafb

Mafb is a basic domain leucine zipper transcription factor (Eichmann et al., 1997) involved in hindbrain segmentation and anterior–posterior patterning (Barrow et al., 2000; Marin and Charnay, 2000). In the mouse, zebrafish, and chicken embryo, Mafb is expressed in r5 and r6 and associated NCCs (Eichmann et al., 1997; Grapin-Botton et al., 1998; Marin and Charnay, 2000) and acts with Krox20, Hoxa1, Hoxa2, and Hoxb2 to pattern these cells (Barrow et al., 2000; McGonnell et al., 2001). Mafb is important for subdivision of the presumptive r5 and r6 territory into definitive rhombomeres and plays an important role in regulating *Hox* gene transcription (Cooke et al., 2001; Prince et al., 1998).

Homozygous *Mafb^{kreisler}* mutant embryos have defects in hindbrain segmentation and inner ear development (Eichmann et al., 1997). In chickens and mice, Mafb is expressed in non-NC tissues, such as vestibular and acoustic nuclei, spinal cord and brain stem neurons, and the mesonephros, but Mafb expression is not lost in some of these tissues in the *Mafb^{kreisler}* homozygous mice, suggesting this allele is not a null (Eichmann et al., 1997). The zebrafish loss-of-function *Mafb^{valentine}* mutant has inappropriately specified NCCs (Prince et al., 1998) and a mixed-identity region one rhombomere in length between r4 and r7. Normally, Ephb4a is expressed with Mafb in r5 and r6, whereas the ligand Efnb2a is expressed on either side, in r4 and r7. In *Mafb^{valentine}* loss-of-function mutant embryos, Ephb4a expression is downregulated and Efnb2a expression is upregulated between r4 and r7, suggesting Mafb sets up mutually exclusive expression domains, establishing the boundaries of the r5/r6 region (Cooke et al., 2001).

Transplantation experiments in chicken embryos demonstrate that signals from somites 7–10 are important for Mafb expression in the hindbrain and corresponding NCCs. Mafb expression is maintained when r5 and r6 are grafted into the r3 and r4 region but repressed when grafted into the r7 and r8 region (Grapin-Botton et al., 1998). RA beads mimic the effect produced by the somites in repressing Mafb in r5/6 and progressively inducing it more rostrally as RA concentration increases (Grapin-Botton et al., 1998). RA signaling, in part by restricting Mafb and Krox20 expression domains, is essential for specifying rhombomere identity and caudal hindbrain segmentation (Dupe et al., 1999). Inactivation of RARα and RARβ results in expansion of the Mafb expression domain to twice normal size, enlargement of r5, disappearance of the r5/r6 boundary, and profound alterations of rhombomere identities (Dupe et al., 1999). *Raldh2*-null mice, which have RA deficiency and die at midgestation, have a severe reduction of Mafb expression (Niederreither et al., 2000). In chicken embryos, application of exogenous FGFs to the NT leads to ectopic expression of Mafb and Krox20 in the NC of the somitic hindbrain (r7 and r8), whereas inhibition of FGF signaling causes downregulation of Mafb and Krox20 (Marin and Charnay, 2000).

15.7 MEDIATOR SUBUNITS Med12 AND Med24

Mediator is a coactivator complex that assists the interaction of DNA-binding transcription factors with RNA polymerase II (Rau et al., 2006). Mediator subunit 12 [Med12, also known as thyroid hormone receptor-associated protein 230 (Trap230)] and Med24 (also known as Trap100) are both important during NC development. In zebrafish, the embryonic lethal homozygous *Med12^{kohtalo}* mutant embryos show defects in brain, NC, and kidney development. In these mutants, cells initiate differentiation pathways in the affected tissues, and many cell type-specific genes are expressed, but differentiation is often incomplete, associated with a failure in morphogenesis (Hong et al., 2005).

Another Med12 mutation was identified in zebrafish from a screen for mutants resembling Sox9 loss-of-function. Med12 acts as a critical coactivator for Sox9, which directs development of NC, otic placodes, cartilage, and bone (Rau et al., 2006).

During development, Med24 is expressed in a pattern consistent with functions in both ENS and craniofacial skeletal development. The *Med24*^{lessen} loss-of-function mutant or knockdown of Med24 has a significant reduction in ENS neurons and defects in cranial NC-derived structures. Initial specification and migration of the NC is unaffected in *Med24*^{lessen} mutants, but proliferation of ENS progenitors is significantly reduced. Med24 is not required for initial steps of cranial NC development but is essential for later proliferation of ENS progenitors. Med24 acts cell autonomously in pharyngeal endoderm and indirectly influences the development of NC-derived cartilages. The intestinal endoderm is also essential for ENS development, and it is unclear if this may be where Med24 is functioning to affect ENS development (Pietsch et al., 2006).

15.8 Mef2

The four Myocyte enhancer factor 2 (Mef2) genes, Mef2a, Mef2b, Mef2c, and Mef2d, encode transcription factors belonging to the MADS (MCM1-agamous-deficiens-serum response factor) family of DNA binding proteins that also includes Srf (Lyons et al., 1995). Mef2 transcription factors bind a conserved A/T-rich sequence in many skeletal and cardiac muscle gene promoters and are some of the earliest markers for the cardiac muscle lineage. Mef2a, Mef2c, and Mef2d are expressed first in the myocardium and NCCs (E8.5–E10.5) and then later in myotomes, prospective limb muscle, and the CNS (Edmondson et al., 1994; Lyons et al., 1995). Reduced expression of Mef2c and Hand1 in *Hif1α-null* embryos may cause defective ventricle formation, and these mutants also have NCC migration defects leading to abnormal OFT and vessel remodeling and hypoplastic pharyngeal arches (Compernolle et al., 2003). Mef2c also interacts with Myocardin (Myocd) to regulate smooth muscle transcription of the *Myocd* gene (Creemers et al., 2006).

In the mouse, Mef2c is required for expression of Dlx5, Dlx6, and Hand2 in the pharyngeal arches. NC-specific deletion of *Mef2c* results in neonatal lethality due to severe craniofacial defects. Mice heterozygous for *Mef2c*, *Dlx5*, or *Dlx6* have no defects, but double heterozygous *Mef2c;Dlx5* mice exhibit defective palate development and neonatal lethality (Verzi et al., 2007). In zebrafish, Mef2c (mef2ca) is also required in cranial NC for proper pharyngeal skeletal patterning; Mef2c mutants have craniofacial defects resembling Edn1 partial loss-of-function and Mef2c interacts genetically with Edn1. Expression of Edn1-dependent target genes *Hand2*, *Dlx4*, *Dlx5*, *Dlx6*, *Bap*, and *Gsc* requires Mef2c function (Miller et al., 2007).

15.9 Mrf2

During differentiation of a multipotent NCC line (MONC-1) toward a VSMC fate, the transcription factors modulator recognition factor 2α and β (Mrf2α and Mrf2β), members of the AT-rich interaction domain (ARID) family of transcription factors known to regulate differentiation, are highly induced. In vivo, Mrf2α is expressed in adult mouse cardiac and vascular tissues. Overexpression of Mrf2α and Mrf2β in 3T3 fibroblasts induces expression of vascular smooth muscle marker genes like *smooth muscle α-actin* and *smooth muscle 22α* and slows proliferation (Watanabe et al., 2002b).

15.10 Myb AND Mybl2

The myeloblastosis protooncogene *Myb* (or *c-Myb*) and the homologous *Mybl2* (*Myb-like 2*, also known as *B-Myb*) encode transcriptional regulators containing a unique Myb domain similar to a helix–turn–helix domain of eukaryotic homeodomain proteins. In the chicken embryo, Myb is expressed in neuroectoderm and participates in the regulation of trunk NCCs. Reduction of endogenous Myb in vitro and in ovo prevents formation of migratory NCCs. Moderate overexpression of Myb in naive intermediate neural plate triggers EMT and NCC migration, probably through cooperation with Bmp4 signaling. Bmp4 activates Myb expression, causing accumulation of transcripts of the Bmp4 downstream genes *Msx1* and *Slug*, important in the early NC. Reduction of Myb prevents Bmp4-induced NC formation (Karafiat et al., 2005). Conditional expression of Myb in an NC-derived neuroblastoma cell line causes upregulation of Igf1, Igf1r, and Igfbp5 expression. The *Igfbp5* promoter has two Myb binding sites bound by Myb and Mybl2 in vitro and in vivo, and these Myb proteins directly enhance transcription from an *Igfbp5* reporter (Tanno et al., 2002). During *Xenopus* embryogenesis, Mybl2 is preferentially expressed in the developing nervous system and NCCs. Within the developing NT, *Xenopus Mybl2* gene transcription occurs preferentially in proliferating, non-differentiated cells (Humbert-Lan and Pieler, 1999).

15.11 Myocd, Mkl1, Mkl2

Myocardin (Myocd) and the Myocd-related transcription factors Mkl1 [megakaryoblastic leukemia/myocardin-like 1, also known as Myocardin-related transcription factor a (Mrtfa)] and Mkl2 (Mrtfb) act as coactivators for Srf and play a key role in cardiovascular development (Li et al., 2005; Oh et al., 2005). Embryos homozygous for an *Mkl2* loss-of-function mutation die perinatally from cardiac OFT defects and thin-walled myocardium, accompanied by an earlier failure in differentiation of VSMCs within the NC-derived pharyngeal arch arteries (Li et al., 2005; Oh et al., 2005). The *Mkl2*-null phenotype is distinct from the *Myocd*-null phenotype, suggesting unique roles for each of these Srf coactivators in development of different subsets of VSMCs in vivo (Oh et al., 2005), despite normal migration and initial patterning of cardiac NCCs (Li et al., 2005). Myocd is the earliest known marker specific to both cardiac and smooth muscle lineages during embryogenesis. The activity of a *Myocd* enhancer in the cardiovascular system requires both Mef2 and FoxO proteins, and Myocd regulates its own enhancer through Mef2 independently of Srf (Creemers et al., 2006).

15.12 Nfatc3 AND Nfatc4

Nuclear factor of activated T-cells (Nfat) transcription factors are a calcium-activated family originally identified by their role in immune response, but several of these are also expressed and function in NCCs. Nfat activity, together with calcineurin, is essential for neuregulin and ErbB signaling and plays a role in NC diversification and differentiation of Schwann cells (Kao et al., 2009). Nfatc3 and Nfatc4 are activated downstream of Calcineurin, which is activated by an increase in cytoplasmic calcium ions upon addition of neuregulin to Schwann cell precursors. Nfatc4 interacts synergistically with Sox10, an Nfat nuclear partner, to activate *Krox20*, which in turn regulates genes necessary for myelination. Mouse embryos lacking CalcineurinB1 in the developing NC have defects in Schwann cell differentiation and myelination (Kao et al., 2009). In undifferentiated cells from the NC-derived PC12 pheochromocytoma cell line, an Nfat-activated reporter construct is transcribed upon addition of exogenous Wnt1 and Wnt7a. However, there may be other factors interpreting signals downstream of Nfat; heterologous expression of Wnt1 enhances proliferation, but opposite effects were observed in PC12 cells expressing Wnt7a (Spinsanti et al., 2008).

15.13 Nfyb

Zebrafish Nfyb (nuclear transcription factor y-β) becomes restricted to prospective craniofacial cartilage regions and the developing notochord. Nfyb knockdown causes reduction in head size, loss of some craniofacial cartilages, and increased apoptosis in the head, with a depletion of cranial NCCs required for mandibular and pharyngeal arch formation (Chen et al., 2009).

15.14 NUCLEAR RECEPTORS Nr2f1, Nr2f2, Nr5a1, Nr5a2

Nr2f (also known as *COUP-TF*) genes encode orphan members of the steroid/thyroid hormone receptor superfamily highly expressed in the vertebrate developing nervous system with putative roles in neuronal development and differentiation. In the mouse, there are two homologous *Nr2f* genes (*Nr2f1 and Nr2f2*), and their expression patterns overlap extensively. *Nr2f1-null* animals die perinatally. Mutant embryos have altered morphogenesis of the ninth cranial ganglion, possibly due to extra cell death in the neuronal precursor cell population. At midgestation, aberrant nerve projection and branching were observed in several other regions of mutant embryos. Nr2f1 is required for proper fetal and postnatal development and has some functions distinct from Nr2f2, which has not been directly investigated in the NC (Qiu et al., 1997).

Nr5a1 (also known as *steroidogenic factor 1* or *Sf1*) heterozygous mutant embryos have a decrease in adrenal precursors starting at E10, but after E13.5, increased cell proliferation in *Nr5a1* heterozygous embryos compensates and almost normal adrenocortical size is restored. NC-derived adrenomedullary precursors migrate normally in heterozygous and null embryos, but later in development, medullary growth is compromised in both genotypes. Despite the small adrenal size in *Nr5a1* heterozygotes, steroidogenic capacity per cell is elevated in their primary adult adrenocortical cells, consistent with upregulation of some Nr5a1 target genes in heterozygous *Nr5a1* adrenal glands (Bland et al., 2004). Expression of Nr5a2, also known as fetoprotein transcription factor (Ftf), occurs in the yolk sac endoderm, pharyngeal arches and NCCs, and in the foregut endoderm during liver and pancreatic morphogenesis (Rausa et al., 1999).

15.15 RELA (NFκB SUBUNIT)

In *Xenopus*, indirect activation of NFκB by injection of *Bcl2l1* mRNA enables NFκB to directly upregulate Slug and Snail, rescuing the mutant phenotype of Slug, a factor normally required for expression of mesodermal and NC markers. Slug indirectly upregulates NFκB subunits and directly downregulates the proapoptotic Caspase9 (Zhang et al., 2006a). Regulation of NFκB activity in rat DRGs and the NC-derived pheochromocytoma cell line PC12 indicates a role for NFκB in delayed rather than immediate-early responses of the PNS and related cell lines to inflammatory cytokines (Wood, 1995). Wnt1 production and NFκB activation are linked in the PC12 cell line (Sun et al., 2008). The neurosteroid dehydroepiandrosterone (DHEA) protects NC-derived PC12 cells from serum deprivation-induced apoptosis. DHEA induces sequential phosphorylation of prosurvival kinases, which then activate Creb1 and NFκB which in turn induce expression of the antiapoptotic *Bcl2* genes (Charalampopoulos et al., 2008).

15.16 Srf

Like Mef2, Serum response factor (Srf) is a member of the MADS family of transcription factors and is a key regulator of the expression of smooth muscle and cardiac muscle genes (Creemers et al., 2006). Cardiac-specific response to Edn1 requires activity of Srf combined with the tissue-restricted Gata proteins. When cotransfected, Srf and Gata factors synergistically activate cardiac-specific Edn1-inducible promoters containing Gata and Srf binding sites (Morin et al., 2001). Srf is a transcriptional coactivator with Myocd; most of the Myocd target genes are dependent on Srf (Creemers et al., 2006). In *Xenopus* animal cap explants containing NCCs, Myocd-dependent induction of smooth muscle genes is synergistically activated by Srf but antagonized by Gata6 (Barillot et al., 2008). Inactivation of Srf, a downstream effector of the Erk cascade, produces conotruncal and craniofacial defects much like those of Erk2 deletion mutants (Newbern et al., 2008). Srf also directly regulates miR-145 and miR-143, microRNAs that promote differentiaton and repress proliferation of SMCs (including those derived from the NC) (Cordes et al., 2009).

15.17 IMMEDIATE-EARLY GENES Sp1, Sp3, AND Sp4

The *Sp* genes are among the immediate-early transcription factors and bind to many promoters. Because of their more ubiquitous nature, transcription mediated by these factors is very likely to be modulated by interactions with other transcription factors. Several motifs for Sp1 binding are present in the regulatory region of the cell surface glycoprotein Mcam, a cell adhesion molecule associated with tumor progression and metastasis in human malignant melanoma (Sers et al., 1993). Transcription from the *Tyr* promoter, able to drive expression of a reporter gene in immortalized quail NCCs, may also be mediated via an Sp1-binding motif (Ferguson and Kidson, 1996). Expression from the *Ret* promoter in the NC-derived TT cell line requires only 70 bp of sequence upstream of the transcription start site, which contains two Sp1/Sp3 binding sites, and expression was abrogated by removal of these Sp1/Sp3 binding sites (Andrew et al., 2000). *Tlx1* transfection enhances the activity of the *Ret* promoter in the SK-N-MC cell line by stimulating a region of the promoter with binding sites for Sp1, altering the interaction of Sp1 with the *Ret* promoter (Bachetti et al., 2005). Overexpression of Gata3 causes an increase in dopamine β-hydroxylase (Dbh)-expressing neurons in primary NCSC culture through two upstream enhancers that do not contain Gata3 binding sites but do contain binding sites for Sp1 and AP4, both of which physically interact with Gata3 (Hong et al., 2008b). In an NC-derived cell line that exhibits the characteristics of immature Schwann cells, Smads may act indirectly to promote GFAP expression through induction of Sp1, which is induced by Bmp2 (Dore et al., 2009). Expression of Sp4 (HF1b) is required for specification of the cardiac conduction system, and tissue-specific deletion of Sp4 shows its requirement in both the cardiomyogenic and NC lineages. Absence of Sp4 in the NC leads to arrhythmogenesis, conduction system defects, and atrial and atrioventricular dysfunction resulting from deficiencies in Ntrk3, a neurotrophin receptor (St Amand et al., 2003; St Amand et al., 2006).

15.18 Stat3

The *signal transducer and activator of transcription 3 (Stat3)* gene encodes a transcription factor that is a marker of pluripotency and one of a cohort of pluripotency genes expressed in NCSCs such as multipotent skin-derived precursors (SKPs) derived from facial skin (Zhao et al., 2009). *Xenopus* Hes4, essential for NC progenitor survival and maintenance, induces Delta1 through Stat3 (Nichane et al., 2008). A conserved Stat binding site provides a major contribution to the expression of AP-2α in the facial prominences (Donner and Williams, 2006). Glial differentiation occurs in response to Bmp4 and is specifically blocked by a dominant-negative Stat3. Upon Bmp4 treatment, Mtor associates with Stat3 and facilitates its activation. Inhibition of Mtor prevents Stat3 activation and glial differentiation (Rajan et al., 2003). In vitro, Cntf and Lif induce nuclear translocation of Stat3 in NC-derived cells, increase neuronal or glial marker expression, and decrease expression of the NC progenitor cell marker, Nestin (Chalazonitis et al., 1998). Dysfunctional signaling in Ret mutants often results in constitutive activation of Stat3 (Plaza Menacho et al., 2005).

15.19 Tcof1

Treacher Collins syndrome is an autosomal-dominant craniofacial disorder caused by haploinsufficiency of the *Tcof1* gene product Treacle (Dixon and Dixon, 2004; Shows and Shiang, 2008). Tcof1 expression is high in developing NC, but much lower in other tissues, and heterozygous knockout of Tcof1 in the mouse embryo causes craniofacial malformation and lethality; this phenotype is very sensitive to genetic background and can be rescued by reducing p53 and inhibiting apoptosis (Dixon and Dixon, 2004; Shows and Shiang, 2008).

15.20 Tead2

The transcriptional enhancer activator domain family member 2 or Tead-box 2 (Tead2) transcription factor, along with its coactivator Yap1, is coexpressed with Pax3 in the dorsal NT. Tead2 binds an essential binding site within a Pax3 NC-specific enhancer to activate Pax3 expression. A Tead2-Engrailed fusion protein represses RA-induced Pax3 expression in vivo and in P19 cells (Milewski et al., 2004). *Tead2*-null mice have a significantly increased risk of NT closure defects, but early expression of Pax3 is normal, suggesting a Pax3-independent early role. NTDs could be suppressed by folic acid or the p53 inhibitor pifithrin-α (Kaneko et al., 2007). Tead2 is also required later specifically for enhancer activation in NC-derived smooth muscle cells and the dorsal aorta (Creemers et al., 2006)

15.21 Tsc22d1

TGFβ1 Stimulated Clone-22 domain 1 (Tsc22d1) is a leucine zipper transcription factor with ubiquitous expression during early development. Tsc22d1 expression is later upregulated mainly at sites of epithelial–mesenchymal interactions (limb bud, tooth primordiurn, hair follicle, kidney, lung, pancreas) and in many NC-derived tissues including pharyngeal arch mesenchyme, cranial ganglia, sympathetic ganglia, DRGs, and the craniofacial skeleton. It is also expressed in the heart and in osteochondrogenic regions throughout the embryo (Dohrmann et al., 1999). Specific function of this protein has not yet been evaluated in the NC.

• • • •

References

Abate, C., and Curran, T. (1990). Encounters with Fos and Jun on the road to AP-1. *Semin Cancer Biol* **1**, pp. 19–26.

Acosta, S., Lavarino, C., Paris, R., Garcia, I., de Torres, C., Rodriguez, E., Beleta, H., and Mora, J. (2009). Comprehensive characterization of neuroblastoma cell line subtypes reveals bilineage potential similar to neural crest stem cells. *BMC Dev Biol* **9**, p. 12.

Addo-Yobo, S. O., Straessle, J., Anwar, A., Donson, A. M., Kleinschmidt-Demasters, B. K., and Foreman, N. K. (2006). Paired overexpression of ErbB3 and Sox10 in pilocytic astrocytoma. *J Neuropathol Exp Neurol* **65**, pp. 769–75.

Ai, D., Liu, W., Ma, L., Dong, F., Lu, M. F., Wang, D., Verzi, M. P., Cai, C., Gage, P. J., Evans, S., Black, B. L., Brown, N. A., and Martin, J. F. (2006). Pitx2 regulates cardiac left–right asymmetry by patterning second cardiac lineage-derived myocardium. *Dev Biol* **296**, pp. 437–49.

Airik, R., Karner, M., Karis, A., and Karner, J. (2005). Gene expression analysis of Gata3-/- mice by using cDNA microarray technology. *Life Sci* **76**, pp. 2559–68.

Akiyama, H., Kim, J. E., Nakashima, K., Balmes, G., Iwai, N., Deng, J. M., Zhang, Z., Martin, J. F., Behringer, R. R., Nakamura, T., and de Crombrugghe, B. (2005). Osteochondroprogenitor cells are derived from Sox9 expressing precursors. *Proc Natl Acad Sci U S A* **102**, pp. 14665–70.

Aldskogius, H., Berens, C., Kanaykina, N., Liakhovitskaia, A., Medvinsky, A., Sandelin, M., Schreiner, S., Wegner, M., Hjerling-Leffler, J., and Kozlova, E. N. (2009). Regulation of boundary cap neural crest stem cell differentiation after transplantation. *Stem Cells* **27**, pp. 1592–603.

Amano, O., Bringas, P., Takahashi, I., Takahashi, K., Yamane, A., Chai, Y., Nuckolls, G. H., Shum, L., and Slavkin, H. C. (1999). Nerve growth factor (NGF) supports tooth morphogenesis in mouse first branchial arch explants. *Dev Dyn* **216**, pp. 299–310.

Anderson, D. J. (1993). MASH genes and the logic of neural crest cell lineage diversification. *C R Acad Sci III* **316**, pp. 1082–96.

Anderson, D. J. (1994). Stem cells and transcription factors in the development of the mammalian neural crest. *Faseb J* **8**, pp. 707–13.

Anderson, D. J., Groves, A., Lo, L., Ma, Q., Rao, M., Shah, N. M., and Sommer, L. (1997). Cell lineage determination and the control of neuronal identity in the neural crest. *Cold Spring Harb Symp Quant Biol* **62**, pp. 493–504.

Andrew, S. D., Delhanty, P. J., Mulligan, L. M., and Robinson, B. G. (2000). Sp1 and Sp3 transactivate the RET proto-oncogene promoter. *Gene* **256**, pp. 283–91.

Angelo, S., Lohr, J., Lee, K. H., Ticho, B. S., Breitbart, R. E., Hill, S., Yost, H. J., and Srivastava, D. (2000). Conservation of sequence and expression of *Xenopus* and zebrafish dHAND during cardiac, branchial arch and lateral mesoderm development. *Mech Dev* **95**, pp. 231–7.

Antonellis, A., Bennett, W. R., Menheniott, T. R., Prasad, A. B., Lee-Lin, S. Q., Green, E. D., Paisley, D., Kelsh, R. N., Pavan, W. J., and Ward, A. (2006). Deletion of long-range sequences at Sox10 compromises developmental expression in a mouse model of Waardenburg-Shah (WS4) syndrome. *Hum Mol Genet* **15**, pp. 259–71.

Antonellis, A., Huynh, J. L., Lee-Lin, S. Q., Vinton, R. M., Renaud, G., Loftus, S. K., Elliot, G., Wolfsberg, T. G., Green, E. D., McCallion, A. S., and Pavan, W. J. (2008). Identification of neural crest and glial enhancers at the mouse Sox10 locus through transgenesis in zebrafish. *PLoS Genet* **4**, p. e1000174.

Aoki, Y., Saint-Germain, N., Gyda, M., Magner-Fink, E., Lee, Y. H., Credidio, C., and Saint-Jeannet, J. P. (2003). Sox10 regulates the development of neural crest-derived melanocytes in *Xenopus*. *Dev Biol* **259**, pp. 19–33.

Aquino, J. B., Hjerling-Leffler, J., Koltzenburg, M., Edlund, T., Villar, M. J., and Ernfors, P. (2006). In vitro and in vivo differentiation of boundary cap neural crest stem cells into mature Schwann cells. *Exp Neurol* **198**, pp. 438–49.

Aragno, M., Mastrocola, R., Medana, C., Catalano, M. G., Vercellinatto, I., Danni, O., and Boccuzzi, G. (2006). Oxidative stress-dependent impairment of cardiac-specific transcription factors in experimental diabetes. *Endocrinology* **147**, pp. 5967–74.

Artinger, K. B., Fedtsova, N., Rhee, J. M., Bronner-Fraser, M., and Turner, E. (1998). Placodal origin of Brn-3-expressing cranial sensory neurons. *J Neurobiol* **36**, pp. 572–85.

Arvidsson, Y., Sumantran, V., Watt, F., Uramoto, H., and Funa, K. (2005). Neuroblastoma-specific cytotoxicity mediated by the Mash1-promoter and *E. coli* purine nucleoside phosphorylase. *Pediatr Blood Cancer* **44**, pp. 77–84.

Aybar, M. J., Nieto, M. A., and Mayor, R. (2003). Snail precedes slug in the genetic cascade required for the specification and migration of the *Xenopus* neural crest. *Development* **130**, pp. 483–94.

Bachetti, T., Borghini, S., Ravazzolo, R., and Ceccherini, I. (2005). An in vitro approach to test the possible role of candidate factors in the transcriptional regulation of the RET proto-oncogene. *Gene Expr* **12**, pp. 137–49.

Bacon, W., Tschill, P., Grollemund, B., Matern, O., Rinkenbach, R., Sauvage, P., Kaufmann, I., Bousquet, P., Brandt, C., and Perrin-Schmitt, F. (2007). [Genetic origin of non-syndromic cleft lip and palate. TWIST, a candidate gene? Research protocol]. *Orthod Fr* **78**, pp. 249–55.

Baker, C. V., and Bronner-Fraser, M. (1997). The origins of the neural crest. Part II: an evolutionary perspective. *Mech Dev* **69**, pp. 13–29.

Baker, C. V., Stark, M. R., and Bronner-Fraser, M. (2002). Pax3-expressing trigeminal placode cells can localize to trunk neural crest sites but are committed to a cutaneous sensory neuron fate. *Dev Biol* **249**, pp. 219–36.

Ball, D. W., Azzoli, C. G., Baylin, S. B., Chi, D., Dou, S., Donis-Keller, H., Cumaraswamy, A., Borges, M., and Nelkin, B. D. (1993). Identification of a human achaete-scute homolog highly expressed in neuroendocrine tumors. *Proc Natl Acad Sci U S A* **90**, pp. 5648–52.

Balling, R., Mutter, G., Gruss, P., and Kessel, M. (1989). Craniofacial abnormalities induced by ectopic expression of the homeobox gene Hox-1.1 in transgenic mice. *Cell* **58**, pp. 337–47.

Baltzinger, M., Mager-Heckel, A. M., and Remy, P. (1999). Xl erg: expression pattern and overexpression during development plead for a role in endothelial cell differentiation. *Dev Dyn* **216**, pp. 420–33.

Baltzinger, M., Ori, M., Pasqualetti, M., Nardi, I., and Rijli, F. M. (2005). Hoxa2 knockdown in *Xenopus* results in hyoid to mandibular homeosis. *Dev Dyn* **234**, pp. 858–67.

Bamforth, S. D., Braganca, J., Eloranta, J. J., Murdoch, J. N., Marques, F. I., Kranc, K. R., Farza, H., Henderson, D. J., Hurst, H. C., and Bhattacharya, S. (2001). Cardiac malformations, adrenal agenesis, neural crest defects and exencephaly in mice lacking Cited2, a new Tfap2 co-activator. *Nat Genet* **29**, pp. 469–74.

Bang, A. G., Papalopulu, N., Goulding, M. D., and Kintner, C. (1999). Expression of Pax-3 in the lateral neural plate is dependent on a Wnt-mediated signal from posterior nonaxial mesoderm. *Dev Biol* **212**, pp. 366–80.

Barbosa, A. C., Funato, N., Chapman, S., McKee, M. D., Richardson, J. A., Olson, E. N., and Yanagisawa, H. (2007). Hand transcription factors cooperatively regulate development of the distal midline mesenchyme. *Dev Biol* **310**, pp. 154–68.

Barembaum, M., and Bronner-Fraser, M. (2005). Early steps in neural crest specification. *Semin Cell Dev Biol* **16**, pp. 642–6.

Barillot, W., Treguer, K., Faucheux, C., Fedou, S., Theze, N., and Thiebaud, P. (2008). Induction and modulation of smooth muscle differentiation in *Xenopus* embryonic cells. *Dev Dyn* **237**, pp. 3373–86.

Barrallo-Gimeno, A., Holzschuh, J., Driever, W., and Knapik, E. W. (2004). Neural crest survival and differentiation in zebrafish depends on mont blanc/tfap2a gene function. *Development* **131**, pp. 1463–77.

Barrow, J. R., and Capecchi, M. R. (1999). Compensatory defects associated with mutations in Hoxa1 restore normal palatogenesis to Hoxa2 mutants. *Development* **126**, pp. 5011–26.

Barrow, J. R., Stadler, H. S., and Capecchi, M. R. (2000). Roles of Hoxa1 and Hoxa2 in patterning the early hindbrain of the mouse. *Development* **127**, pp. 933–44.

Basch, M. L., and Bronner-Fraser, M. (2006). Neural crest inducing signals. *Adv Exp Med Biol* **589**, pp. 24–31.

Basch, M. L., Bronner-Fraser, M., and Garcia-Castro, M. I. (2006). Specification of the neural crest occurs during gastrulation and requires Pax7. *Nature* **441**, pp. 218–22.

Basch, M. L., Selleck, M. A., and Bronner-Fraser, M. (2000). Timing and competence of neural crest formation. *Dev Neurosci* **22**, pp. 217–27.

Bastidas, F., De Calisto, J., and Mayor, R. (2004). Identification of neural crest competence territory: role of Wnt signaling. *Dev Dyn* **229**, pp. 109–17.

Bates, M. D., Dunagan, D. T., Welch, L. C., Kaul, A., and Harvey, R. P. (2006). The Hlx homeobox transcription factor is required early in enteric nervous system development. *BMC Dev Biol* **6**, p. 33.

Baulmann, D. C., Ohlmann, A., Flugel-Koch, C., Goswami, S., Cvekl, A., and Tamm, E. R. (2002). Pax6 heterozygous eyes show defects in chamber angle differentiation that are associated with a wide spectrum of other anterior eye segment abnormalities. *Mech Dev* **118**, pp. 3–17.

Baxter, L. L., and Pavan, W. J. (2002). The oculocutaneous albinism type IV gene Matp is a new marker of pigment cell precursors during mouse embryonic development. *Mech Dev* **116**, pp. 209–12.

Baxter, L. L., and Pavan, W. J. (2003). Pmel17 expression is Mitf-dependent and reveals cranial melanoblast migration during murine development. *Gene Expr Patterns* **3**, pp. 703–7.

Begemann, G., Schilling, T. F., Rauch, G. J., Geisler, R., and Ingham, P. W. (2001). The zebrafish neckless mutation reveals a requirement for raldh2 in mesodermal signals that pattern the hindbrain. *Development* **128**, pp. 3081–94.

Bejar, J., Hong, Y., and Schartl, M. (2003). Mitf expression is sufficient to direct differentiation of medaka blastula derived stem cells to melanocytes. *Development* **130**, pp. 6545–53.

Bellmeyer, A., Krase, J., Lindgren, J., and LaBonne, C. (2003). The protooncogene c-myc is an essential regulator of neural crest formation in xenopus. *Dev Cell* **4**, pp. 827–39.

Bel-Vialar, S., Itasaki, N., and Krumlauf, R. (2002). Initiating Hox gene expression: in the early chick neural tube differential sensitivity to FGF and RA signaling subdivides the HoxB genes in two distinct groups. *Development* **129**, pp. 5103–15.

Benailly, H. K., Lapierre, J. M., Laudier, B., Amiel, J., Attie, T., De Blois, M. C., Vekemans, M., and Romana, S. P. (2003). PMX2B, a new candidate gene for Hirschsprung's disease. *Clin Genet* **64**, pp. 204–9.

Benjanirut, C., Paris, M., Wang, W. H., Hong, S. J., Kim, K. S., Hullinger, R. L., and Andrisani, O. M. (2006). The cAMP pathway in combination with BMP2 regulates Phox2a transcription via cAMP response element binding sites. *J Biol Chem* **281**, pp. 2969–81.

Bennetts, J. S., Rendtorff, N. D., Simpson, F., Tranebjaerg, L., and Wicking, C. (2007). The coding region of TP53INP2, a gene expressed in the developing nervous system, is not altered in a family with autosomal recessive non-progressive infantile ataxia on chromosome 20q11-q13. *Dev Dyn* **236**, pp. 843–52.

Berdal, A., Lezot, F., Nefussi, J. R., and Sautier, J. M. (2000). [Mineralized dental tissues: a unique example of skeletal biodiversity derived from cephaic neural crest]. *Morphologie* **84**, pp. 5–10.

Beverdam, A., Brouwer, A., Reijnen, M., Korving, J., and Meijlink, F. (2001). Severe nasal clefting and abnormal embryonic apoptosis in Alx3/Alx4 double mutant mice. *Development* **128**, pp. 3975–86.

Bhatheja, K., and Field, J. (2006). Schwann cells: origins and role in axonal maintenance and regeneration. *Int J Biochem Cell Biol* **38**, pp. 1995–9.

Bilodeau, M. L., Boulineau, T., Greulich, J. D., Hullinger, R. L., and Andrisani, O. M. (2001). Differential expression of sympathoadrenal lineage-determining genes and phenotypic markers in cultured primary neural crest cells. *In Vitro Cell Dev Biol Anim* **37**, pp. 185–92.

Bilodeau, M. L., Boulineau, T., Hullinger, R. L., and Andrisani, O. M. (2000). Cyclic AMP signaling functions as a bimodal switch in sympathoadrenal cell development in cultured primary neural crest cells. *Mol Cell Biol* **20**, pp. 3004–14.

Blanco, M. J., Moreno-Bueno, G., Sarrio, D., Locascio, A., Cano, A., Palacios, J., and Nieto, M. A. (2002). Correlation of Snail expression with histological grade and lymph node status in breast carcinomas. *Oncogene* **21**, pp. 3241–6.

Bland, M. L., Fowkes, R. C., and Ingraham, H. A. (2004). Differential requirement for steroidogenic factor-1 gene dosage in adrenal development versus endocrine function. *Mol Endocrinol* **18**, pp. 941–52.

Boissy, R. E., and Nordlund, J. J. (1997). Molecular basis of congenital hypopigmentary disorders in humans: a review. *Pigment Cell Res* **10**, pp. 12–24.

Bondurand, N., Kobetz, A., Pingault, V., Lemort, N., Encha-Razavi, F., Couly, G., Goerich, D. E., Wegner, M., Abitbol, M., and Goossens, M. (1998). Expression of the SOX10 gene during human development. *FEBS Lett* **432**, pp. 168–72.

Bondurand, N., Natarajan, D., Barlow, A., Thapar, N., and Pachnis, V. (2006). Maintenance of mammalian enteric nervous system progenitors by SOX10 and endothelin 3 signalling. *Development* **133**, pp. 2075–86.

Bordogna, W., Hudson, J. D., Buddle, J., Bennett, D. C., Beach, D. H., and Carnero, A. (2005).

EMX homeobox genes regulate microphthalmia and alter melanocyte biology. *Exp Cell Res* **311**, pp. 27–38.

Bothe, I., and Dietrich, S. (2006). The molecular setup of the avian head mesoderm and its implication for craniofacial myogenesis. *Dev Dyn* **235**, pp. 2845–60.

Borghini, S., Bachetti, T., Fava, M., Di Duca, M., Cargnin, F., Fornasari, D., Ravazzolo, R., and Ceccherini, I. (2006). The TLX2 homeobox gene is a transcriptional target of PHOX2B in neural-crest-derived cells. *Biochem J* **395**, pp. 355–61.

Bowles, J., Schepers, G., and Koopman, P. (2000). Phylogeny of the SOX family of developmental transcription factors based on sequence and structural indicators. *Dev Biol* **227**, pp. 239–55.

Bradley, L. C., Snape, A., Bhatt, S., and Wilkinson, D. G. (1993). The structure and expression of the *Xenopus* Krox-20 gene: conserved and divergent patterns of expression in rhombomeres and neural crest. *Mech Dev* **40**, pp. 73–84.

Braganca, J., Swingler, T., Marques, F. I., Jones, T., Eloranta, J. J., Hurst, H. C., Shioda, T., and Bhattacharya, S. (2002). Human CREB-binding protein/p300-interacting transactivator with ED-rich tail (CITED) 4, a new member of the CITED family, functions as a co-activator for transcription factor AP-2. *J Biol Chem* **277**, pp. 8559–65.

Brandl, C., Florian, C., Driemel, O., Weber, B. H., and Morsczeck, C. (2009). Identification of neural crest-derived stem cell-like cells from the corneal limbus of juvenile mice. *Exp Eye Res* **89**, pp. 209–17.

Brewer, A., Nemer, G., Gove, C., Rawlins, F., Nemer, M., Patient, R., and Pizzey, J. (2002a). Widespread expression of an extended peptide sequence of GATA-6 during murine embryogenesis and non-equivalence of RNA and protein expression domains. *Mech Dev* **119 Suppl 1**, pp. S121–9.

Brewer, A., Nemer, G., Gove, C., Rawlins, F., Nemer, M., Patient, R., and Pizzey, J. (2002b). Widespread expression of an extended peptide sequence of GATA-6 during murine embryogenesis and non-equivalence of RNA and protein expression domains. *Gene Expr Patterns* **2**, pp. 123–31.

Brewer, S., Feng, W., Huang, J., Sullivan, S., and Williams, T. (2004). Wnt1-Cre-mediated deletion of AP-2alpha causes multiple neural crest-related defects. *Dev Biol* **267**, pp. 135–52.

Brewer, S., Jiang, X., Donaldson, S., Williams, T., and Sucov, H. M. (2002c). Requirement for AP-2alpha in cardiac outflow tract morphogenesis. *Mech Dev* **110**, pp. 139–49.

Brewster, R., Lee, J., and Ruiz i Altaba, A. (1998). Gli/Zic factors pattern the neural plate by defining domains of cell differentiation. *Nature* **393**, pp. 579–83.

Briones, V. R., Chen, S., Riegel, A. T., and Lechleider, R. J. (2006). Mechanism of fibroblast growth factor-binding protein 1 repression by TGF-beta. *Biochem Biophys Res Commun* **345**, pp. 595–601.

Britsch, S., Goerich, D. E., Riethmacher, D., Peirano, R. I., Rossner, M., Nave, K. A., Birchmeier, C., and Wegner, M. (2001). The transcription factor Sox10 is a key regulator of peripheral glial development. *Genes Dev* **15**, pp. 66–78.

Brizzolara, A., Torre, M., Favre, A., Pini Prato, A., Bocciardi, R., and Martucciello, G. (2004). Histochemical study of Dom mouse: A model for Waardenburg–Hirschsprung's phenotype. *J Pediatr Surg* **39**, pp. 1098–103.

Broders-Bondon, F., Chesneau, A., Romero-Oliva, F., Mazabraud, A., Mayor, R., and Thiery, J. P. (2007). Regulation of XSnail2 expression by Rho GTPases. *Dev Dyn* **236**, pp. 2555–66.

Brown, L. A., Amores, A., Schilling, T. F., Jowett, T., Baert, J. L., de Launoit, Y., and Sharrocks, A. D. (1998). Molecular characterization of the zebrafish PEA3 ETS-domain transcription factor. *Oncogene* **17**, pp. 93–104.

Buac, K., Watkins-Chow, D. E., Loftus, S. K., Larson, D. M., Incao, A., Gibney, G., and Pavan, W. J. (2008). A Sox10 expression screen identifies an amino acid essential for Erbb3 function. *PLoS Genet* **4**, p. e1000177.

Burns, A. J., and Delalande, J. M. (2005). Neural crest cell origin for intrinsic ganglia of the developing chicken lung. *Dev Biol* **277**, pp. 63–79.

Burstyn-Cohen, T., Stanleigh, J., Sela-Donenfeld, D., and Kalcheim, C. (2004). Canonical Wnt activity regulates trunk neural crest delamination linking BMP/noggin signaling with G1/S transition. *Development* **131**, pp. 5327–39.

Buxton, P., Hunt, P., Ferretti, P., and Thorogood, P. (1997). A role for midline closure in the re-establishment of dorsoventral pattern following dorsal hindbrain ablation. *Dev Biol* **183**, pp. 150–65.

Cacheux, V., Dastot-Le Moal, F., Kaariainen, H., Bondurand, N., Rintala, R., Boissier, B., Wilson, M., Mowat, D., and Goossens, M. (2001). Loss-of-function mutations in SIP1 Smad interacting protein 1 result in a syndromic Hirschsprung disease. *Hum Mol Genet* **10**, pp. 1503–10.

Cairns, L. A., Crotta, S., Minuzzo, M., Ricciardi-Castagnoli, P., Pozzi, L., and Ottolenghi, S. (1997). Immortalization of neuro-endocrine cells from adrenal tumors arising in SV40 T-transgenic mice. *Oncogene* **14**, pp. 3093–8.

Cantrell, V. A., Owens, S. E., Chandler, R. L., Airey, D. C., Bradley, K. M., Smith, J. R., and Southard-Smith, E. M. (2004). Interactions between Sox10 and EdnrB modulate penetrance and severity of aganglionosis in the *Sox10^{Dom}* mouse model of Hirschsprung disease. *Hum Mol Genet* **13**, pp. 2289–301.

Carl, T. F., Dufton, C., Hanken, J., and Klymkowsky, M. W. (1999). Inhibition of neural crest migration in *Xenopus* using antisense slug RNA. *Dev Biol* **213**, pp. 101–15.

Carney, T. J., Dutton, K. A., Greenhill, E., Delfino-Machin, M., Dufourcq, P., Blader, P., and

Kelsh, R. N. (2006). A direct role for Sox10 in specification of neural crest-derived sensory neurons. *Development* **133**, pp. 4619–30.

Cebra-Thomas, J. A., Betters, E., Yin, M., Plafkin, C., McDow, K., and Gilbert, S. F. (2007). Evidence that a late-emerging population of trunk neural crest cells forms the plastron bones in the turtle Trachemys scripta. *Evol Dev* **9**, pp. 267–77.

Chai, Y., Ito, Y., and Han, J. (2003). TGF-beta signaling and its functional significance in regulating the fate of cranial neural crest cells. *Crit Rev Oral Biol Med* **14**, pp. 78–88.

Chalazonitis, A., Rothman, T. P., Chen, J., Vinson, E. N., MacLennan, A. J., and Gershon, M. D. (1998). Promotion of the development of enteric neurons and glia by neuropoietic cytokines: interactions with neurotrophin-3. *Dev Biol* **198**, pp. 343–65.

Chang, C. P., Stankunas, K., Shang, C., Kao, S. C., Twu, K. Y., and Cleary, M. L. (2008). Pbx1 functions in distinct regulatory networks to pattern the great arteries and cardiac outflow tract. *Development* **135**, pp. 3577–86.

Chappell, J. H., Jr., Wang, X. D., and Loeken, M. R. (2009). Diabetes and apoptosis: neural crest cells and neural tube. *Apoptosis* **14**, pp. 1472–83.

Charalampopoulos, I., Margioris, A. N., and Gravanis, A. (2008). Neurosteroid dehydroepiandrosterone exerts anti-apoptotic effects by membrane-mediated, integrated genomic and non-genomic pro-survival signaling pathways. *J Neurochem* **107**, pp. 1457–69.

Charite, J., McFadden, D. G., Merlo, G., Levi, G., Clouthier, D. E., Yanagisawa, M., Richardson, J. A., and Olson, E. N. (2001). Role of Dlx6 in regulation of an endothelin-1-dependent, dHAND branchial arch enhancer. *Genes Dev* **15**, pp. 3039–49.

Chazaud, C., Oulad-Abdelghani, M., Bouillet, P., Decimo, D., Chambon, P., and Dolle, P. (1996). AP-2.2, a novel gene related to AP-2, is expressed in the forebrain, limbs and face during mouse embryogenesis. *Mech Dev* **54**, pp. 83–94.

Chelyshev Iu, A., and Saitkulov, K. I. (2000). [The development, phenotypic characteristics and communications of Schwann cells]. *Usp Fiziol Nauk* **31**, pp. 54–69.

Chen, J., Han, M., Manisastry, S. M., Trotta, P., Serrano, M. C., Huhta, J. C., and Linask, K. K. (2008). Molecular effects of lithium exposure during mouse and chick gastrulation and subsequent valve dysmorphogenesis. *Birth Defects Res A Clin Mol Teratol* **82**, pp. 508–18.

Chen, S., Ji, M., Paris, M., Hullinger, R. L., and Andrisani, O. M. (2005). The cAMP pathway regulates both transcription and activity of the paired homeobox transcription factor Phox2a required for development of neural crest-derived and central nervous system-derived catecholaminergic neurons. *J Biol Chem* **280**, pp. 41025–36.

Chen, S., Kulik, M., and Lechleider, R. J. (2003). Smad proteins regulate transcriptional induction of the SM22alpha gene by TGF-beta. *Nucleic Acids Res* **31**, pp. 1302–10.

Chen, Y. H., Ishii, M., Sun, J., Sucov, H. M., and Maxson, R. E., Jr. (2007). Msx1 and Msx2 regulate survival of secondary heart field precursors and post-migratory proliferation of cardiac neural crest in the outflow tract. *Dev Biol* **308**, pp. 421–37.

Chen, Y. H., Layne, M. D., Watanabe, M., Yet, S. F., and Perrella, M. A. (2001). Upstream stimulatory factors regulate aortic preferentially expressed gene-1 expression in vascular smooth muscle cells. *J Biol Chem* **276**, pp. 47658–63.

Chen, Y. H., Lin, Y. T., and Lee, G. H. (2009). Novel and unexpected functions of zebrafish CCAAT box binding transcription factor (NF-Y) B subunit during cartilages development. *Bone* **44**, pp. 777–84.

Chen, Z. F., and Behringer, R. R. (1995). twist is required in head mesenchyme for cranial neural tube morphogenesis. *Genes Dev* **9**, pp. 686–99.

Cheng, Y., Cheung, M., Abu-Elmagd, M. M., Orme, A., and Scotting, P. J. (2000). Chick sox10, a transcription factor expressed in both early neural crest cells and central nervous system. *Brain Res Dev Brain Res* **121**, pp. 233–41.

Cheung, M., and Briscoe, J. (2003). Neural crest development is regulated by the transcription factor Sox9. *Development* **130**, pp. 5681–93.

Cheung, M., Chaboissier, M. C., Mynett, A., Hirst, E., Schedl, A., and Briscoe, J. (2005). The transcriptional control of trunk neural crest induction, survival, and delamination. *Dev Cell* **8**, pp. 179–92.

Chhin, B., Hatayama, M., Bozon, D., Ogawa, M., Schon, P., Tohmonda, T., Sassolas, F., Aruga, J., Valard, A. G., Chen, S. C., and Bouvagnet, P. (2007). Elucidation of penetrance variability of a ZIC3 mutation in a family with complex heart defects and functional analysis of ZIC3 mutations in the first zinc finger domain. *Hum Mutat* **28**, pp. 563–70.

Chiba, S., Kurokawa, M. S., Yoshikawa, H., Ikeda, R., Takeno, M., Tadokoro, M., Sekino, H., Hashimoto, T., and Suzuki, N. (2005). Noggin and basic FGF were implicated in forebrain fate and caudal fate, respectively, of the neural tube-like structures emerging in mouse ES cell culture. *Exp Brain Res* **163**, pp. 86–99.

Choi, Y. J., Yoon, T. J., and Lee, Y. H. (2008). Changing expression of the genes related to human hair graying. *Eur J Dermatol* **18**, pp. 397–9.

Chomette, D., Frain, M., Cereghini, S., Charnay, P., and Ghislain, J. (2006). Krox20 hindbrain cis-regulatory landscape: interplay between multiple long-range initiation and autoregulatory elements. *Development* **133**, pp. 1253–62.

Cinatl, J., Jr., Cinatl, J., Vogel, J. U., Rabenau, H., Kornhuber, B., and Doerr, H. W. (1996). Modulatory effects of human cytomegalovirus infection on malignant properties of cancer cells. *Intervirology* **39**, pp. 259–69.

Clark, M. S., Lanigan, T. M., Page, N. M., and Russo, A. F. (1995). Induction of a serotonergic and neuronal phenotype in thyroid C-cells. *J Neurosci* **15**, pp. 6167–78.

Clouthier, D. E., Hosoda, K., Richardson, J. A., Williams, S. C., Yanagisawa, H., Kuwaki, T., Kumada, M., Hammer, R. E., and Yanagisawa, M. (1998). Cranial and cardiac neural crest defects in endothelin-A receptor-deficient mice. *Development* **125**, pp. 813–24.

Clouthier, D. E., Williams, S. C., Yanagisawa, H., Wieduwilt, M., Richardson, J. A., and Yanagisawa, M. (2000). Signaling pathways crucial for craniofacial development revealed by endothelin-A receptor-deficient mice. *Dev Biol* **217**, pp. 10–24.

Cobourne, M. T. (2000). Construction for the modern head: current concepts in craniofacial development. *J Orthod* **27**, pp. 307–14.

Coles, E. G., Lawlor, E. R., and Bronner-Fraser, M. (2008). EWS-FLI1 causes neuroepithelial defects and abrogates emigration of neural crest stem cells. *Stem Cells* **26**, pp. 2237–44.

Coles, E. G., Taneyhill, L. A., and Bronner-Fraser, M. (2007). A critical role for Cadherin6B in regulating avian neural crest emigration. *Dev Biol* **312**, pp. 533–44.

Collinson, J. M., Quinn, J. C., Hill, R. E., and West, J. D. (2003). The roles of Pax6 in the cornea, retina, and olfactory epithelium of the developing mouse embryo. *Dev Biol* **255**, pp. 303–12.

Combs, M. D., and Yutzey, K. E. (2009). Heart valve development: regulatory networks in development and disease. *Circ Res* **105**, pp. 408–21.

Compernolle, V., Brusselmans, K., Franco, D., Moorman, A., Dewerchin, M., Collen, D., and Carmeliet, P. (2003). Cardia bifida, defective heart development and abnormal neural crest migration in embryos lacking hypoxia-inducible factor-1alpha. *Cardiovasc Res* **60**, pp. 569–79.

Conway, S. J., Bundy, J., Chen, J., Dickman, E., Rogers, R., and Will, B. M. (2000). Decreased neural crest stem cell expansion is responsible for the conotruncal heart defects within the splotch (Sp(2H))/Pax3 mouse mutant. *Cardiovasc Res* **47**, pp. 314–28.

Conway, S. J., Godt, R. E., Hatcher, C. J., Leatherbury, L., Zolotouchnikov, V. V., Brotto, M. A., Copp, A. J., Kirby, M. L., and Creazzo, T. L. (1997a). Neural crest is involved in development of abnormal myocardial function. *J Mol Cell Cardiol* **29**, pp. 2675–85.

Conway, S. J., Henderson, D. J., and Copp, A. J. (1997b). Pax3 is required for cardiac neural crest migration in the mouse: evidence from the splotch (Sp2H) mutant. *Development* **124**, pp. 505–14.

Conway, S. J., Henderson, D. J., Kirby, M. L., Anderson, R. H., and Copp, A. J. (1997c). Development of a lethal congenital heart defect in the splotch (Pax3) mutant mouse. *Cardiovasc Res* **36**, pp. 163–73.

Cook, A. L., Donatien, P. D., Smith, A. G., Murphy, M., Jones, M. K., Herlyn, M., Bennett, D. C., Leonard, J. H., and Sturm, R. A. (2003). Human melanoblasts in culture: expression of BRN2 and synergistic regulation by fibroblast growth factor-2, stem cell factor, and endothelin-3. *J Invest Dermatol* **121**, pp. 1150–9.

Cooke, J., Moens, C., Roth, L., Durbin, L., Shiomi, K., Brennan, C., Kimmel, C., Wilson, S., and Holder, N. (2001). Eph signalling functions downstream of Val to regulate cell sorting and boundary formation in the caudal hindbrain. *Development* **128**, pp. 571–80.

Cordes, K. R., Sheehy, N. T., White, M. P., Berry, E. C., Morton, S. U., Muth, A. N., Lee, T. H., Miano, J. M., Ivey, K. N., and Srivastava, D. (2009). miR-145 and miR-143 regulate smooth muscle cell fate and plasticity. *Nature* **460**, pp. 705–10.

Cornish, E. J., Hassan, S. M., Martin, J. D., Li, S., and Merzdorf, C. S. (2009). A microarray screen for direct targets of Zic1 identifies an aquaporin gene, aqp-3b, expressed in the neural folds. *Dev Dyn* **238**, pp. 1179–94.

Corry, G. N., Hendzel, M. J., and Underhill, D. A. (2008). Subnuclear localization and mobility are key indicators of PAX3 dysfunction in Waardenburg syndrome. *Hum Mol Genet* **17**, pp. 1825–37.

Corry, G. N., and Underhill, D. A. (2005). Pax3 target gene recognition occurs through distinct modes that are differentially affected by disease-associated mutations. *Pigment Cell Res* **18**, pp. 427–38.

Coura, G. S., Garcez, R. C., de Aguiar, C. B., Alvarez-Silva, M., Magini, R. S., and Trentin, A. G. (2008). Human periodontal ligament: a niche of neural crest stem cells. *J Periodontal Res* **43**, pp. 531–6.

Creemers, E. E., Sutherland, L. B., McAnally, J., Richardson, J. A., and Olson, E. N. (2006). Myocardin is a direct transcriptional target of Mef2, Tead and Foxo proteins during cardiovascular development. *Development* **133**, pp. 4245–56.

Cserjesi, P., Brown, D., Lyons, G. E., and Olson, E. N. (1995). Expression of the novel basic helix–loop–helix gene eHAND in neural crest derivatives and extraembryonic membranes during mouse development. *Dev Biol* **170**, pp. 664–78.

Cuesta, I., Zaret, K. S., and Santisteban, P. (2007). The forkhead factor FoxE1 binds to the thyroperoxidase promoter during thyroid cell differentiation and modifies compacted chromatin structure. *Mol Cell Biol* **27**, pp. 7302–14.

Curran, K., Raible, D. W., and Lister, J. A. (2009). Foxd3 controls melanophore specification in the zebrafish neural crest by regulation of Mitf. *Dev Biol* **332**, pp. 408–17.

Cvekl, A., and Tamm, E. R. (2004). Anterior eye development and ocular mesenchyme: new insights from mouse models and human diseases. *Bioessays* **26**, pp. 374–86.

Cvekl, A., and Wang, W. L. (2009). Retinoic acid signaling in mammalian eye development. *Exp Eye Res* **89**, pp. 280–91.

Dai, Y. S., and Cserjesi, P. (2002). The basic helix–loop–helix factor, HAND2, functions as a transcriptional activator by binding to E-boxes as a heterodimer. *J Biol Chem* **277**, pp. 12604–12.

Dai, Y. S., Hao, J., Bonin, C., Morikawa, Y., and Cserjesi, P. (2004). JAB1 enhances HAND2 transcriptional activity by regulating HAND2 DNA binding. *J Neurosci Res* **76**, pp. 613–22.

Damberg, M., Westberg, L., Berggard, C., Landen, M., Sundblad, C., Eriksson, O., Naessen, T., Ekman, A., and Eriksson, E. (2005). Investigation of transcription factor AP-2 beta genotype in women with premenstrual dysphoric disorder. *Neurosci Lett* **377**, pp. 49–52.

Dambly-Chaudiere, C., and Vervoort, M. (1998). The bHLH genes in neural development. *Int J Dev Biol* **42**, pp. 269–73.

Dammer, R., Stavenow, J., Held, P., Schroder, J., Niederdellmann, H., Hofstadter, F., and Buettner, R. (1997). Pigmented peripheral nerve sheath tumor of the oral cavity with expression of AP-2 beta and c-RET: a case report. *Oral Surg Oral Med Oral Pathol Oral Radiol Endod* **84**, pp. 40–4.

D'Autreaux, F., Morikawa, Y., Cserjesi, P., and Gershon, M. D. (2007). Hand2 is necessary for terminal differentiation of enteric neurons from crest-derived precursors but not for their migration into the gut or for formation of glia. *Development* **134**, pp. 2237–49.

Davideau, J. L., Demri, P., Gu, T. T., Simmons, D., Nessman, C., Forest, N., MacDougall, M., and Berdal, A. (1999). Expression of DLX5 during human embryonic craniofacial development. *Mech Dev* **81**, pp. 183–6.

Davidson, D. (1995). The function and evolution of Msx genes: pointers and paradoxes. *Trends Genet* **11**, pp. 405–11.

Davidson, L. A., and Keller, R. E. (1999). Neural tube closure in *Xenopus laevis* involves medial migration, directed protrusive activity, cell intercalation and convergent extension. *Development* **126**, pp. 4547–56.

de Jong, R., and Meijlink, F. (1993). The homeobox gene S8: mesoderm-specific expression in presomite embryos and in cells cultured in vitro and modulation in differentiating pluripotent cells. *Dev Biol* **157**, pp. 133–46.

de Pontual, L., Nepote, V., Attie-Bitach, T., Al Halabiah, H., Trang, H., Elghouzzi, V., Levacher, B., Benihoud, K., Auge, J., Faure, C., Laudier, B., Vekemans, M., Munnich, A., Perricaudet, M., Guillemot, F., Gaultier, C., Lyonnet, S., Simonneau, M., and Amiel, J. (2003). Noradrenergic neuronal development is impaired by mutation of the proneural HASH-1 gene in congenital central hypoventilation syndrome (Ondine's curse). *Hum Mol Genet* **12**, pp. 3173–80.

Deal, K. K., Cantrell, V. A., Chandler, R. L., Saunders, T. L., Mortlock, D. P., and Southard-Smith, E. M. (2006). Distant regulatory elements in a Sox10-beta GEO BAC transgene are required for expression of Sox10 in the enteric nervous system and other neural crest-derived tissues. *Dev Dyn* **235**, pp. 1413–32.

Deflorian, G., Tiso, N., Ferretti, E., Meyer, D., Blasi, F., Bortolussi, M., and Argenton, F. (2004). Prep1.1 has essential genetic functions in hindbrain development and cranial neural crest cell differentiation. *Development* **131**, pp. 613–27.

Degistirici, O., Jaquiery, C., Schonebeck, B., Siemonsmeier, J., Gotz, W., Martin, I., and Thie, M. (2008). Defining properties of neural crest-derived progenitor cells from the apex of human developing tooth. *Tissue Eng Part A* **14**, pp. 317–30.

del Barrio, M. G., and Nieto, M. A. (2002). Overexpression of Snail family members highlights their ability to promote chick neural crest formation. *Development* **129**, pp. 1583–93.

Del Barrio, M. G., and Nieto, M. A. (2004). Relative expression of Slug, RhoB, and HNK-1 in the cranial neural crest of the early chicken embryo. *Dev Dyn* **229**, pp. 136–9.

Deng, Z. L., Sharff, K. A., Tang, N., Song, W. X., Luo, J., Luo, X., Chen, J., Bennett, E., Reid, R., Manning, D., Xue, A., Montag, A. G., Luu, H. H., Haydon, R. C., and He, T. C. (2008). Regulation of osteogenic differentiation during skeletal development. *Front Biosci* **13**, pp. 2001–21.

Dhordain, P., Dewitte, F., Desbiens, X., Stehelin, D., and Duterque-Coquillaud, M. (1995). Mesodermal expression of the chicken erg gene associated with precartilaginous condensation and cartilage differentiation. *Mech Dev* **50**, pp. 17–28.

Dick, A., Mayr, T., Bauer, H., Meier, A., and Hammerschmidt, M. (2000). Cloning and characterization of zebrafish smad2, smad3 and smad4. *Gene* **246**, pp. 69–80.

Dickinson, M. E., Selleck, M. A., McMahon, A. P., and Bronner-Fraser, M. (1995). Dorsalization of the neural tube by the non-neural ectoderm. *Development* **121**, pp. 2099–106.

Dim, D. C., Cooley, L. D., and Miranda, R. N. (2007). Clear cell sarcoma of tendons and aponeuroses: a review. *Arch Pathol Lab Med* **131**, pp. 152–6.

Dixon, J., and Dixon, M. J. (2004). Genetic background has a major effect on the penetrance and severity of craniofacial defects in mice heterozygous for the gene encoding the nucleolar protein Treacle. *Dev Dyn* **229**, pp. 907–14.

Dohrmann, C. E., Belaoussoff, M., and Raftery, L. A. (1999). Dynamic expression of TSC-22 at sites of epithelial–mesenchymal interactions during mouse development. *Mech Dev* **84**, pp. 147–51.

Donnell, A. M., Bannigan, J., and Puri, P. (2005). The effect of vagal neural crest ablation on the chick embryo cloaca. *Pediatr Surg Int* **21**, pp. 180–3.

Donner, A. L., and Williams, T. (2006). Frontal nasal prominence expression driven by Tcfap2a relies on a conserved binding site for STAT proteins. *Dev Dyn* **235**, pp. 1358–70.

Dore, J. J., DeWitt, J. C., Setty, N., Donald, M. D., Joo, E., Chesarone, M. A., and Birren, S. J. (2009). Multiple signaling pathways converge to regulate bone-morphogenetic-protein-dependent glial gene expression. *Dev Neurosci* **31**, pp. 473–86.

Dorsky, R. I., Raible, D. W., and Moon, R. T. (2000). Direct regulation of nacre, a zebrafish MITF homolog required for pigment cell formation, by the Wnt pathway. *Genes Dev* **14**, pp. 158–62.

Dottori, M., Gross, M. K., Labosky, P., and Goulding, M. (2001). The winged-helix transcription factor Foxd3 suppresses interneuron differentiation and promotes neural crest cell fate. *Development* **128**, pp. 4127–38.

Drerup, C. M., Wiora, H. M., Topczewski, J., and Morris, J. A. (2009). Disc1 regulates foxd3 and sox10 expression, affecting neural crest migration and differentiation. *Development* **136**, pp. 2623–32.

Duband, J. L., Monier, F., Delannet, M., and Newgreen, D. (1995). Epithelium-mesenchyme transition during neural crest development. *Acta Anat (Basel)* **154**, pp. 63–78.

Dubreuil, V., Hirsch, M. R., Jouve, C., Brunet, J. F., and Goridis, C. (2002). The role of Phox2b in synchronizing pan-neuronal and type-specific aspects of neurogenesis. *Development* **129**, pp. 5241–53.

Dupe, V., Ghyselinck, N. B., Wendling, O., Chambon, P., and Mark, M. (1999). Key roles of retinoic acid receptors alpha and beta in the patterning of the caudal hindbrain, pharyngeal arches and otocyst in the mouse. *Development* **126**, pp. 5051–9.

Dupin, E., Calloni, G., Real, C., Goncalves-Trentin, A., and Le Douarin, N. M. (2007). Neural crest progenitors and stem cells. *C R Biol* **330**, pp. 521–9.

Dupin, E., Creuzet, S., and Le Douarin, N. M. (2006). The contribution of the neural crest to the vertebrate body. *Adv Exp Med Biol* **589**, pp. 96–119.

Dutton, J. R., Antonellis, A., Carney, T. J., Rodrigues, F. S., Pavan, W. J., Ward, A., and Kelsh, R. N. (2008). An evolutionarily conserved intronic region controls the spatiotemporal expression of the transcription factor Sox10. *BMC Dev Biol* **8**, p. 105.

Dutton, K. A., Pauliny, A., Lopes, S. S., Elworthy, S., Carney, T. J., Rauch, J., Geisler, R., Haffter, P., and Kelsh, R. N. (2001). Zebrafish colourless encodes sox10 and specifies non-ectomesenchymal neural crest fates. *Development* **128**, pp. 4113–25.

Dy, P., Han, Y., and Lefebvre, V. (2008). Generation of mice harboring a Sox5 conditional null allele. *Genesis* **46**, pp. 294–9.

Ebert, S. N., Ficklin, M. B., Her, S., Siddall, B. J., Bell, R. A., Ganguly, K., Morita, K., and Wong, D. L. (1998). Glucocorticoid-dependent action of neural crest factor AP-2: stimulation of phenylethanolamine *N*-methyltransferase gene expression. *J Neurochem* **70**, pp. 2286–95.

Edmondson, D. G., Lyons, G. E., Martin, J. F., and Olson, E. N. (1994). Mef2 gene expression marks the cardiac and skeletal muscle lineages during mouse embryogenesis. *Development* **120**, pp. 1251–63.

Eferl, R., Sibilia, M., Hilberg, F., Fuchsbichler, A., Kufferath, I., Guertl, B., Zenz, R., Wagner, E. F., and Zatloukal, K. (1999). Functions of c-Jun in liver and heart development. *J Cell Biol* **145**, pp. 1049–61.

Eichmann, A., Grapin-Botton, A., Kelly, L., Graf, T., Le Douarin, N. M., and Sieweke, M. (1997). The expression pattern of the mafB/kr gene in birds and mice reveals that the kreisler phenotype does not represent a null mutant. *Mech Dev* **65**, pp. 111–22.

Eisen, T. G. (1996). The control of gene expression in melanocytes and melanomas. *Melanoma Res* **6**, pp. 277–84.

Eliazer, S., Spencer, J., Ye, D., Olson, E., and Ilaria, R. L., Jr. (2003). Alteration of mesodermal cell differentiation by EWS/FLI-1, the oncogene implicated in Ewing's sarcoma. *Mol Cell Biol* **23**, pp. 482–92.

Ellies, D. L., and Krumlauf, R. (2006). Bone formation: The nuclear matrix reloaded. *Cell* **125**, pp. 840–2.

Ellies, D. L., Langille, R. M., Martin, C. C., Akimenko, M. A., and Ekker, M. (1997). Specific craniofacial cartilage dysmorphogenesis coincides with a loss of dlx gene expression in retinoic acid-treated zebrafish embryos. *Mech Dev* **61**, pp. 23–36.

Elms, P., Siggers, P., Napper, D., Greenfield, A., and Arkell, R. (2003). Zic2 is required for neural crest formation and hindbrain patterning during mouse development. *Dev Biol* **264**, pp. 391–406.

Elworthy, S., Lister, J. A., Carney, T. J., Raible, D. W., and Kelsh, R. N. (2003). Transcriptional regulation of mitfa accounts for the sox10 requirement in zebrafish melanophore development. *Development* **130**, pp. 2809–18.

Elworthy, S., Pinto, J. P., Pettifer, A., Cancela, M. L., and Kelsh, R. N. (2005). Phox2b function in the enteric nervous system is conserved in zebrafish and is sox10-dependent. *Mech Dev* **122**, pp. 659–69.

Endo, Y., Osumi, N., and Wakamatsu, Y. (2003). Deltex/Dtx mediates NOTCH signaling in regulation of Bmp4 expression in cranial neural crest formation during avian development. *Dev Growth Differ* **45**, pp. 241–8.

Eng, S. R., Dykes, I. M., Lanier, J., Fedtsova, N., and Turner, E. E. (2007). POU-domain factor Brn3a regulates both distinct and common programs of gene expression in the spinal and trigeminal sensory ganglia. *Neural Dev* **2**, p. 3.

Engleka, K. A., Gitler, A. D., Zhang, M., Zhou, D. D., High, F. A., and Epstein, J. A. (2005). Insertion of Cre into the Pax3 locus creates a new allele of Splotch and identifies unexpected Pax3 derivatives. *Dev Biol* **280**, pp. 396–406.

Epperlein, H., Meulemans, D., Bronner-Fraser, M., Steinbeisser, H., and Selleck, M. A. (2000). Analysis of cranial neural crest migratory pathways in axolotl using cell markers and transplantation. *Development* **127**, pp. 2751–61.

Epstein, J. A., Li, J., Lang, D., Chen, F., Brown, C. B., Jin, F., Lu, M. M., Thomas, M., Liu, E., Wessels, A., and Lo, C. W. (2000). Migration of cardiac neural crest cells in Splotch embryos. *Development* **127**, pp. 1869–78.

Eroglu, B., Wang, G., Tu, N., Sun, X., and Mivechi, N. F. (2006). Critical role of Brg1 member of the SWI/SNF chromatin remodeling complex during neurogenesis and neural crest induction in zebrafish. *Dev Dyn* **235**, pp. 2722–35.

Evans, A. L., and Gage, P. J. (2005). Expression of the homeobox gene Pitx2 in neural crest is required for optic stalk and ocular anterior segment development. *Hum Mol Genet* **14**, pp. 3347–59.

Fafeur, V., Tulasne, D., Queva, C., Vercamer, C., Dimster, V., Mattot, V., Stehelin, D., Desbiens, X., and Vandenbunder, B. (1997). The ETS1 transcription factor is expressed during epithelial–mesenchymal transitions in the chick embryo and is activated in scatter factor-stimulated MDCK epithelial cells. *Cell Growth Differ* **8**, pp. 655–65.

Feijoo, C. G., Saldias, M. P., De la Paz, J. F., Gomez-Skarmeta, J. L., and Allende, M. L. (2009). Formation of posterior cranial placode derivatives requires the Iroquois transcription factor irx4a. *Mol Cell Neurosci* **40**, pp. 328–37.

Fenby, B. T., Fotaki, V., and Mason, J. O. (2008). Pax3 regulates Wnt1 expression via a conserved binding site in the 5' proximal promoter. *Biochim Biophys Acta* **1779**, pp. 115–21.

Fendrich, V., Waldmann, J., Feldmann, G., Schlosser, K., Konig, A., Ramaswamy, A., Bartsch, D. K., and Karakas, E. (2009). Unique expression pattern of the EMT markers Snail, Twist and E-cadherin in benign and malignant parathyroid neoplasia. *Eur J Endocrinol* **160**, pp. 695–703.

Ferguson, C. A., and Kidson, S. H. (1996). Characteristic sequences in the promoter region of the chicken tyrosinase-encoding gene. *Gene* **169**, pp. 191–95.

Ferguson, C. A., Tucker, A. S., and Sharpe, P. T. (2000). Temporospatial cell interactions regulating mandibular and maxillary arch patterning. *Development* **127**, pp. 403–12.

Fernandes, K. J., and Miller, F. D. (2009). Isolation, expansion, and differentiation of mouse skin-derived precursors. *Methods Mol Biol* **482**, pp. 159–70.

Filippi, A., Tiso, N., Deflorian, G., Zecchin, E., Bortolussi, M., and Argenton, F. (2005). The basic helix–loop–helix olig3 establishes the neural plate boundary of the trunk and is necessary for development of the dorsal spinal cord. *Proc Natl Acad Sci U S A* **102**, pp. 4377–82.

Firulli, A. B., and Conway, S. J. (2004). Combinatorial transcriptional interaction within the cardiac neural crest: a pair of HANDs in heart formation. *Birth Defects Res C Embryo Today* **72**, pp. 151–61.

Firulli, A. B., and Conway, S. J. (2008). Phosphoregulation of Twist1 provides a mechanism of cell fate control. *Curr Med Chem* **15**, pp. 2641–7.

Firulli, A. B., McFadden, D. G., Lin, Q., Srivastava, D., and Olson, E. N. (1998). Heart and extra-embryonic mesodermal defects in mouse embryos lacking the bHLH transcription factor Hand1. *Nat Genet* **18**, pp. 266–70.

Floris, G., Debiec-Rychter, M., Wozniak, A., Magrini, E., Manfioletti, G., De Wever, I., Tallini, G., and Sciot, R. (2007). Malignant ectomesenchymoma: genetic profile reflects rhabdomyosarcomatous differentiation. *Diagn Mol Pathol* **16**, pp. 243–8.

Foerst-Potts, L., and Sadler, T. W. (1997). Disruption of Msx-1 and Msx-2 reveals roles for these genes in craniofacial, eye, and axial development. *Dev Dyn* **209**, pp. 70–84.

Franco, P. G., Paganelli, A. R., Lopez, S. L., and Carrasco, A. E. (1999). Functional association of retinoic acid and hedgehog signaling in *Xenopus* primary neurogenesis. *Development* **126**, pp. 4257–65.

Franz, T., and Kothary, R. (1993). Characterization of the neural crest defect in Splotch (Sp1H) mutant mice using a lacZ transgene. *Brain Res Dev Brain Res* **72**, pp. 99–105.

Fuchtbauer, E. M. (1995). Expression of M-twist during postimplantation development of the mouse. *Dev Dyn* **204**, pp. 316–22.

Fujimi, T. J., and Aruga, J. (2008). Upstream stimulatory factors, USF1 and USF2 are differentially expressed during *Xenopus* embryonic development. *Gene Expr Patterns* **8**, pp. 376–81.

Fujimi, T. J., Mikoshiba, K., and Aruga, J. (2006). *Xenopus* Zic4: conservation and diversification of expression profiles and protein function among the *Xenopus* Zic family. *Dev Dyn* **235**, pp. 3379–86.

Fukuhara, S., Kurihara, Y., Arima, Y., Yamada, N., and Kurihara, H. (2004). Temporal requirement of signaling cascade involving endothelin-1/endothelin receptor type A in branchial arch development. *Mech Dev* **121**, pp. 1223–33.

Funato, N., Chapman, S. L., McKee, M. D., Funato, H., Morris, J. A., Shelton, J. M., Richardson, J. A., and Yanagisawa, H. (2009). Hand2 controls osteoblast differentiation in the branchial arch by inhibiting DNA binding of Runx2. *Development* **136**, pp. 615–25.

Fuse, N., Yasumoto, K., Takeda, K., Amae, S., Yoshizawa, M., Udono, T., Takahashi, K., Tamai, M., Tomita, Y., Tachibana, M., and Shibahara, S. (1999). Molecular cloning of cDNA encoding a novel microphthalmia-associated transcription factor isoform with a distinct amino-terminus. *J Biochem* **126**, pp. 1043–51.

Gadson, P. F., Jr., Rossignol, C., McCoy, J., and Rosenquist, T. H. (1993). Expression of elastin, smooth muscle alpha-actin, and c-jun as a function of the embryonic lineage of vascular smooth muscle cells. *In Vitro Cell Dev Biol Anim* **29A**, pp. 773–81.

Gajavelli, S., Wood, P. M., Pennica, D., Whittemore, S. R., and Tsoulfas, P. (2004). BMP signaling initiates a neural crest differentiation program in embryonic rat CNS stem cells. *Exp Neurol* **188**, pp. 205–23.

Gale, E., Prince, V., Lumsden, A., Clarke, J., Holder, N., and Maden, M. (1996). Late effects of retinoic acid on neural crest and aspects of rhombomere. *Development* **122**, pp. 783–93.

Galy, A., Neron, B., Planque, N., Saule, S., and Eychene, A. (2002). Activated MAPK/ERK kinase (MEK-1) induces transdifferentiation of pigmented epithelium into neural retina. *Dev Biol* **248**, pp. 251–64.

Gavalas, A., Studer, M., Lumsden, A., Rijli, F. M., Krumlauf, R., and Chambon, P. (1998). Hoxa1 and Hoxb1 synergize in patterning the hindbrain, cranial nerves and second pharyngeal arch. *Development* **125**, pp. 1123–36.

Gavalas, A., Trainor, P., Ariza-McNaughton, L., and Krumlauf, R. (2001). Synergy between Hoxa1 and Hoxb1: the relationship between arch patterning and the generation of cranial neural crest. *Development* **128**, pp. 3017–27.

Geli, J., Nord, B., Frisk, T., Edstrom Elder, E., Ekstrom, T. J., Carling, T., Backdahl, M., and Larsson, C. (2005). Deletions and altered expression of the RIZ1 tumour suppressor gene in 1p36 in pheochromocytomas and abdominal paragangliomas. *Int J Oncol* **26**, pp. 1385–91.

Gelineau-van Waes, J., Smith, L., van Waes, M., Wilberding, J., Eudy, J. D., Bauer, L. K., and Maddox, J. (2008). Altered expression of the iron transporter Nramp1 (Slc11a1) during fetal development of the retinal pigment epithelium in microphthalmia-associated transcription factor Mitf(mi) and Mitf(vitiligo) mouse mutants. *Exp Eye Res* **86**, pp. 419–33.

Georgiades, P., Wood, J., and Brickell, P. M. (1998). Retinoid X receptor-gamma gene expression is developmentally regulated in the embryonic rodent peripheral nervous system. *Anat Embryol (Berl)* **197**, pp. 477–84.

Germanguz, I., Lev, D., Waisman, T., Kim, C. H., and Gitelman, I. (2007). Four twist genes in zebrafish, four expression patterns. *Dev Dyn* **236**, pp. 2615–26.

Gershon, T. R., Oppenheimer, O., Chin, S. S., and Gerald, W. L. (2005). Temporally regulated neural crest transcription factors distinguish neuroectodermal tumors of varying malignancy and differentiation. *Neoplasia* **7**, pp. 575–84.

Gestblom, C., Grynfeld, A., Ora, I., Ortoft, E., Larsson, C., Axelson, H., Sandstedt, B., Cserjesi, P., Olson, E. N., and Pahlman, S. (1999). The basic helix–loop–helix transcription factor dHAND, a marker gene for the developing human sympathetic nervous system, is expressed in both high- and low-stage neuroblastomas. *Lab Invest* **79**, pp. 67–79.

Ghislain, J., Desmarquet-Trin-Dinh, C., Gilardi-Hebenstreit, P., Charnay, P., and Frain, M. (2003). Neural crest patterning: autoregulatory and crest-specific elements co-operate for Krox20 transcriptional control. *Development* **130**, pp. 941–53.

Giannini, G., Di Marcotullio, L., Ristori, E., Zani, M., Crescenzi, M., Scarpa, S., Piaggio, G., Vacca, A., Peverali, F. A., Diana, F., Screpanti, I., Frati, L., and Gulino, A. (1999). HMGI(Y) and

HMGI-C genes are expressed in neuroblastoma cell lines and tumors and affect retinoic acid responsiveness. *Cancer Res* **59**, pp. 2484–92.

Gilmour, D. T., Maischein, H. M., and Nusslein-Volhard, C. (2002). Migration and function of a glial subtype in the vertebrate peripheral nervous system. *Neuron* **34**, pp. 577–88.

Girard, M., and Goossens, M. (2006). Sumoylation of the SOX10 transcription factor regulates its transcriptional activity. *FEBS Lett* **580**, pp. 1635–41.

Gitelman, I. (1997). Twist protein in mouse embryogenesis. *Dev Biol* **189**, pp. 205–14.

Glavic, A., Maris Honore, S., Gloria Feijoo, C., Bastidas, F., Allende, M. L., and Mayor, R. (2004a). Role of BMP signaling and the homeoprotein Iroquois in the specification of the cranial placodal field. *Dev Biol* **272**, pp. 89–103.

Glavic, A., Silva, F., Aybar, M. J., Bastidas, F., and Mayor, R. (2004b). Interplay between Notch signaling and the homeoprotein Xirol is required for neural crest induction in *Xenopus* embryos. *Development* **131**, pp. 347–59.

Glogarova, K., and Buckiova, D. (2004). Changes in sialylation in homozygous Sp2H mouse mutant embryos. *Birth Defects Res A Clin Mol Teratol* **70**, pp. 142–52.

Gomez-Skarmeta, J. L., de la Calle-Mustienes, E., Modolell, J., and Mayor, R. (1999). *Xenopus* brain factor-2 controls mesoderm, forebrain and neural crest development. *Mech Dev* **80**, pp. 15–27.

Gonzalez-Reyes, S., Fernandez-Dumont, V., Calonge, W. M., Martinez, L., and Tovar, J. A. (2006). Vitamin A improves Pax3 expression that is decreased in the heart of rats with experimental diaphragmatic hernia. *J Pediatr Surg* **41**, pp. 327–30.

Gonzalez-Reyes, S., Fernandez-Dumont, V., Martinez-Calonge, W., Martinez, L., Hernandez, F., and Tovar, J. (2005). Pax3 mRNA is decreased in the hearts of rats with experimental diaphragmatic hernia. *Pediatr Surg Int* **21**, pp. 203–7.

Gotoh, M., Izutsu, Y., and Maeno, M. (2003). Complementary expression of AP-2 and AP-2rep in ectodermal derivatives of *Xenopus* embryos. *Dev Genes Evol* **213**, pp. 363–7.

Gottlieb, S., Hanes, S. D., Golden, J. A., Oakey, R. J., and Budarf, M. L. (1998). Goosecoid-like, a gene deleted in DiGeorge and velocardiofacial syndromes, recognizes DNA with a bicoid-like specificity and is expressed in the developing mouse brain. *Hum Mol Genet* **7**, pp. 1497–505.

Gould, D. B., and Walter, M. A. (2000). Cloning, characterization, localization, and mutational screening of the human BARX1 gene. *Genomics* **68**, pp. 336–42.

Goulding, M. D., Chalepakis, G., Deutsch, U., Erselius, J. R., and Gruss, P. (1991). Pax-3, a novel murine DNA binding protein expressed during early neurogenesis. *Embo J* **10**, pp. 1135–47.

Grapin-Botton, A., Bonnin, M. A., Sieweke, M., and Le Douarin, N. M. (1998). Defined concentrations of a posteriorizing signal are critical for MafB/Kreisler segmental expression in the hindbrain. *Development* **125**, pp. 1173–81.

Greenwood, A. L., Turner, E. E., and Anderson, D. J. (1999). Identification of dividing, determined sensory neuron precursors in the mammalian neural crest. *Development* **126**, pp. 3545–59.

Griffith, A. V., Cardenas, K., Carter, C., Gordon, J., Iberg, A., Engleka, K., Epstein, J. A., Manley, N. R., and Richie, E. R. (2009). Increased thymus- and decreased parathyroid-fated organ domains in Splotch mutant embryos. *Dev Biol* **327**, pp. 216–27.

Grimmer, M. R., and Weiss, W. A. (2006). Childhood tumors of the nervous system as disorders of normal development. *Curr Opin Pediatr* **18**, pp. 634–8.

Guemar, L., de Santa Barbara, P., Vignal, E., Maurel, B., Fort, P., and Faure, S. (2007). The small GTPase RhoV is an essential regulator of neural crest induction in *Xenopus*. *Dev Biol* **310**, pp. 113–28.

Guillemot, F., Lo, L. C., Johnson, J. E., Auerbach, A., Anderson, D. J., and Joyner, A. L. (1993). Mammalian achaete-scute homolog 1 is required for the early development of olfactory and autonomic neurons. *Cell* **75**, pp. 463–76.

Gupta, P. B., Kuperwasser, C., Brunet, J. P., Ramaswamy, S., Kuo, W. L., Gray, J. W., Naber, S. P., and Weinberg, R. A. (2005). The melanocyte differentiation program predisposes to metastasis after neoplastic transformation. *Nat Genet* **37**, pp. 1047–54.

Gut, P., Huber, K., Lohr, J., Bruhl, B., Oberle, S., Treier, M., Ernsberger, U., Kalcheim, C., and Unsicker, K. (2005). Lack of an adrenal cortex in Sf1 mutant mice is compatible with the generation and differentiation of chromaffin cells. *Development* **132**, pp. 4611–9.

Haberland, M., Mokalled, M. H., Montgomery, R. L., and Olson, E. N. (2009). Epigenetic control of skull morphogenesis by histone deacetylase 8. *Genes Dev* **23**, pp. 1625–30.

Hagedorn, L., Paratore, C., Brugnoli, G., Baert, J. L., Mercader, N., Suter, U., and Sommer, L. (2000). The Ets domain transcription factor Erm distinguishes rat satellite glia from Schwann cells and is regulated in satellite cells by neuregulin signaling. *Dev Biol* **219**, pp. 44–58.

Hakami, R. M., Hou, L., Baxter, L. L., Loftus, S. K., Southard-Smith, E. M., Incao, A., Cheng, J., and Pavan, W. J. (2006). Genetic evidence does not support direct regulation of EDNRB by SOX10 in migratory neural crest and the melanocyte lineage. *Mech Dev* **123**, pp. 124–34.

Hall, A. K. (1992). Retinoids and a retinoic acid receptor differentially modulate thymosin beta 10 gene expression in transfected neuroblastoma cells. *Cell Mol Neurobiol* **12**, pp. 45–58.

Hall, C., Flores, M. V., Murison, G., Crosier, K., and Crosier, P. (2006). An essential role for zebrafish Fgfrl1 during gill cartilage development. *Mech Dev* **123**, pp. 925–40.

Hammerschmidt, M., and Nusslein-Volhard, C. (1993). The expression of a zebrafish gene homologous to *Drosophila* snail suggests a conserved function in invertebrate and vertebrate gastrulation. *Development* **119**, pp. 1107–18.

Hammond, C. L., Hinits, Y., Osborn, D. P., Minchin, J. E., Tettamanti, G., and Hughes, S. M. (2007). Signals and myogenic regulatory factors restrict pax3 and pax7 expression to dermomyotome-like tissue in zebrafish. *Dev Biol* **302**, pp. 504–21.

Han, J., Ishii, M., Bringas, P., Jr., Maas, R. L., Maxson, R. E., Jr., and Chai, Y. (2007). Concerted action of Msx1 and Msx2 in regulating cranial neural crest cell differentiation during frontal bone development. *Mech Dev* **124**, pp. 729–45.

Harris, M. L., and Erickson, C. A. (2007). Lineage specification in neural crest cell pathfinding. *Dev Dyn* **236**, pp. 1–19.

He, S. J., Stevens, G., Braithwaite, A. W., and Eccles, M. R. (2005). Transfection of melanoma cells with antisense PAX3 oligonucleotides additively complements cisplatin-induced cytotoxicity. *Mol Cancer Ther* **4**, pp. 996–1003.

Heglind, M., Cederberg, A., Aquino, J., Lucas, G., Ernfors, P., and Enerback, S. (2005). Lack of the central nervous system- and neural crest-expressed forkhead gene Foxs1 affects motor function and body weight. *Mol Cell Biol* **25**, pp. 5616–25.

Hendershot, T. J., Liu, H., Clouthier, D. E., Shepherd, I. T., Coppola, E., Studer, M., Firulli, A. B., Pittman, D. L., and Howard, M. J. (2008). Conditional deletion of Hand2 reveals critical functions in neurogenesis and cell type-specific gene expression for development of neural crest-derived noradrenergic sympathetic ganglion neurons. *Dev Biol* **319**, pp. 179–91.

Hendershot, T. J., Liu, H., Sarkar, A. A., Giovannucci, D. R., Clouthier, D. E., Abe, M., and Howard, M. J. (2007). Expression of Hand2 is sufficient for neurogenesis and cell type-specific gene expression in the enteric nervous system. *Dev Dyn* **236**, pp. 93–105.

Henderson, D. J., Ybot-Gonzalez, P., and Copp, A. J. (1997). Over-expression of the chondroitin sulphate proteoglycan versican is associated with defective neural crest migration in the Pax3 mutant mouse (splotch). *Mech Dev* **69**, pp. 39–51.

Herbarth, B., Pingault, V., Bondurand, N., Kuhlbrodt, K., Hermans-Borgmeyer, I., Puliti, A., Lemort, N., Goossens, M., and Wegner, M. (1998). Mutation of the Sry-related Sox10 gene in Dominant megacolon, a mouse model for human Hirschsprung disease. *Proc Natl Acad Sci U S A* **95**, pp. 5161–5.

Hernandez-Lagunas, L., Choi, I. F., Kaji, T., Simpson, P., Hershey, C., Zhou, Y., Zon, L., Mercola, M., and Artinger, K. B. (2005). Zebrafish narrowminded disrupts the transcription factor prdm1 and is required for neural crest and sensory neuron specification. *Dev Biol* **278**, pp. 347–57.

Hirsch, N., and Harris, W. A. (1997). *Xenopus* Brn-3.0, a POU-domain gene expressed in the developing retina and tectum. Not regulated by innervation. *Invest Ophthalmol Vis Sci* **38**, pp. 960–9.

Hoffman, T. L., Javier, A. L., Campeau, S. A., Knight, R. D., and Schilling, T. F. (2007). Tfap2 transcription factors in zebrafish neural crest development and ectodermal evolution. *J Exp Zool B Mol Dev Evol* **308**, pp. 679–91.

Hollenberg, S. M., Sternglanz, R., Cheng, P. F., and Weintraub, H. (1995). Identification of a new family of tissue-specific basic helix–loop–helix proteins with a two-hybrid system. *Mol Cell Biol* **15**, pp. 3813–22.

Holleville, N., Mateos, S., Bontoux, M., Bollerot, K., and Monsoro-Burq, A. H. (2007). Dlx5 drives Runx2 expression and osteogenic differentiation in developing cranial suture mesenchyme. *Dev Biol* **304**, pp. 860–74.

Holzschuh, J., Barrallo-Gimeno, A., Ettl, A. K., Durr, K., Knapik, E. W., and Driever, W. (2003). Noradrenergic neurons in the zebrafish hindbrain are induced by retinoic acid and require tfap2a for expression of the neurotransmitter phenotype. *Development* **130**, pp. 5741–54.

Hong, C. S., and Saint-Jeannet, J. P. (2007). The activity of Pax3 and Zic1 regulates three distinct cell fates at the neural plate border. *Mol Biol Cell* **18**, pp. 2192–202.

Hong, H. K., Lass, J. H., and Chakravarti, A. (1999). Pleiotropic skeletal and ocular phenotypes of the mouse mutation congenital hydrocephalus (ch/Mf1) arise from a winged helix/forkhead transcription factor gene. *Hum Mol Genet* **8**, pp. 625–37.

Hong, S. J., Chae, H., Lardaro, T., Hong, S., and Kim, K. S. (2008a). Trim11 increases expression of dopamine beta-hydroxylase gene by interacting with Phox2b. *Biochem Biophys Res Commun* **368**, pp. 650–5.

Hong, S. J., Choi, H. J., Hong, S., Huh, Y., Chae, H., and Kim, K. S. (2008b). Transcription factor GATA-3 regulates the transcriptional activity of dopamine beta-hydroxylase by interacting with Sp1 and AP4. *Neurochem Res* **33**, pp. 1821–31.

Hong, S. J., Huh, Y., Chae, H., Hong, S., Lardaro, T., and Kim, K. S. (2006). GATA-3 regulates the transcriptional activity of tyrosine hydroxylase by interacting with CREB. *J Neurochem* **98**, pp. 773–81.

Hong, S. J., Lardaro, T., Oh, M. S., Huh, Y., Ding, Y., Kang, U. J., Kirfel, J., Buettner, R., and Kim, K. S. (2008c). Regulation of the noradrenaline neurotransmitter phenotype by the transcription factor AP-2beta. *J Biol Chem* **283**, pp. 16860–7.

Hong, S. K., Haldin, C. E., Lawson, N. D., Weinstein, B. M., Dawid, I. B., and Hukriede, N. A. (2005). The zebrafish kohtalo/trap230 gene is required for the development of the brain, neural crest, and pronephric kidney. *Proc Natl Acad Sci U S A* **102**, pp. 18473–8.

Hong, S. K., Tsang, M., and Dawid, I. B. (2008d). The mych gene is required for neural crest survival during zebrafish development. *PLoS One* **3**, p. e2029.

Honore, S. M., Aybar, M. J., and Mayor, R. (2003). Sox10 is required for the early development of the prospective neural crest in *Xenopus* embryos. *Dev Biol* **260**, pp. 79–96.

Hoover, F., and Glover, J. C. (1998). Regional pattern of retinoid X receptor-alpha gene expression in the central nervous system of the chicken embryo and its up-regulation by exposure to 9-cis retinoic acid. *J Comp Neurol* **398**, pp. 575–86.

Hopwood, N. D., Pluck, A., and Gurdon, J. B. (1989). A *Xenopus* mRNA related to *Drosophila* twist is expressed in response to induction in the mesoderm and the neural crest. *Cell* **59**, pp. 893–903.

Hornyak, T. J., Hayes, D. J., Chiu, L. Y., and Ziff, E. B. (2001). Transcription factors in melanocyte development: distinct roles for Pax-3 and Mitf. *Mech Dev* **101**, pp. 47–59.

Hou, L., Panthier, J. J., and Arnheiter, H. (2000). Signaling and transcriptional regulation in the neural crest-derived melanocyte lineage: interactions between KIT and MITF. *Development* **127**, pp. 5379–89.

Hou, L., and Pavan, W. J. (2008). Transcriptional and signaling regulation in neural crest stem cell-derived melanocyte development: do all roads lead to Mitf? *Cell Res* **18**, pp. 1163–76.

Houston, D. W., and Wylie, C. (2005). Maternal *Xenopus* Zic2 negatively regulates Nodal-related gene expression during anteroposterior patterning. *Development* **132**, pp. 4845–55.

Houzelstein, D., Cohen, A., Buckingham, M. E., and Robert, B. (1997). Insertional mutation of the mouse Msx1 homeobox gene by an nlacZ reporter gene. *Mech Dev* **65**, pp. 123–33.

Howard, M., Foster, D. N., and Cserjesi, P. (1999). Expression of HAND gene products may be sufficient for the differentiation of avian neural crest-derived cells into catecholaminergic neurons in culture. *Dev Biol* **215**, pp. 62–77.

Howard, M. J. (2005). Mechanisms and perspectives on differentiation of autonomic neurons. *Dev Biol* **277**, pp. 271–86.

Howard, M. J., Stanke, M., Schneider, C., Wu, X., and Rohrer, H. (2000). The transcription factor dHAND is a downstream effector of BMPs in sympathetic neuron specification. *Development* **127**, pp. 4073–81.

Hoyt, P. R., Bartholomew, C., Davis, A. J., Yutzey, K., Gamer, L. W., Potter, S. S., Ihle, J. N., and Mucenski, M. L. (1997). The Evi1 proto-oncogene is required at midgestation for neural, heart, and paraxial mesenchyme development. *Mech Dev* **65**, pp. 55–70.

Hromas, R., Ye, H., Spinella, M., Dmitrovsky, E., Xu, D., and Costa, R. H. (1999). Genesis, a Winged Helix transcriptional repressor, has embryonic expression limited to the neural crest, and stimulates proliferation in vitro in a neural development model. *Cell Tissue Res* **297**, pp. 371–82.

Hsieh, Y. W., Zhang, X. M., Lin, E., Oliver, G., and Yang, X. J. (2002). The homeobox gene Six3 is a potential regulator of anterior segment formation in the chick eye. *Dev Biol* **248**, pp. 265–80.

Huang, T., Lin, A. E., Cox, G. F., Golden, W. L., Feldman, G. L., Ute, M., Schrander-Stumpel, C., Kamisago, M., and Vermeulen, S. J. (2002). Cardiac phenotypes in chromosome 4q- syndrome with and without a deletion of the dHAND gene. *Genet Med* **4**, pp. 464–7.

Huber, K., Bruhl, B., Guillemot, F., Olson, E. N., Ernsberger, U., and Unsicker, K. (2002a). Development of chromaffin cells depends on MASH1 function. *Development* **129**, pp. 4729–38.

Huber, K., Combs, S., Ernsberger, U., Kalcheim, C., and Unsicker, K. (2002b). Generation of neuroendocrine chromaffin cells from sympathoadrenal progenitors: beyond the glucocorticoid hypothesis. *Ann N Y Acad Sci* **971**, pp. 554–9.

Huber, K., Karch, N., Ernsberger, U., Goridis, C., and Unsicker, K. (2005). The role of Phox2B in chromaffin cell development. *Dev Biol* **279**, pp. 501–8.

Huber, W. E., Price, E. R., Widlund, H. R., Du, J., Davis, I. J., Wegner, M., and Fisher, D. E. (2003). A tissue-restricted cAMP transcriptional response: SOX10 modulates alpha-melanocyte-stimulating hormone-triggered expression of microphthalmia-associated transcription factor in melanocytes. *J Biol Chem* **278**, pp. 45224–30.

Hudson, C. D., Podesta, J., Henderson, D., Latchman, D. S., and Budhram-Mahadeo, V. (2004). Coexpression of Brn-3a POU protein with p53 in a population of neuronal progenitor cells is associated with differentiation and protection against apoptosis. *J Neurosci Res* **78**, pp. 803–14.

Hudson, C. D., Sayan, A. E., Melino, G., Knight, R. A., Latchman, D. S., and Budhram-Mahadeo, V. (2008). Brn-3a/POU4F1 interacts with and differentially affects p73-mediated transcription. *Cell Death Differ* **15**, pp. 1266–78.

Hu-Lieskovan, S., Zhang, J., Wu, L., Shimada, H., Schofield, D. E., and Triche, T. J. (2005). EWS-FLI1 fusion protein up-regulates critical genes in neural crest development and is responsible for the observed phenotype of Ewing's family of tumors. *Cancer Res* **65**, pp. 4633–44.

Humbert-Lan, G., and Pieler, T. (1999). Regulation of DNA binding activity and nuclear transport of B-Myb in *Xenopus* oocytes. *J Biol Chem* **274**, pp. 10293–300.

Hunter, M. P., and Prince, V. E. (2002). Zebrafish hox paralogue group 2 genes function redundantly as selector genes to pattern the second pharyngeal arch. *Dev Biol* **247**, pp. 367–89.

Ichimiya, S., Nimura, Y., Seki, N., Ozaki, T., Nagase, T., and Nakagawara, A. (2001). Downregulation of hASH1 is associated with the retinoic acid-induced differentiation of human neuroblastoma cell lines. *Med Pediatr Oncol* **36**, pp. 132–4.

Ido, A., and Ito, K. (2006). Expression of chondrogenic potential of mouse trunk neural crest cells by FGF2 treatment. *Dev Dyn* **235**, pp. 361–7.

Ignatius, M. S., Moose, H. E., El-Hodiri, H. M., and Henion, P. D. (2008). colgate/hdac1 Repression of foxd3 expression is required to permit mitfa-dependent melanogenesis. *Dev Biol* **313**, pp. 568–83.

Iida, K., Koseki, H., Kakinuma, H., Kato, N., Mizutani-Koseki, Y., Ohuchi, H., Yoshioka, H., Noji, S., Kawamura, K., Kataoka, Y., Ueno, F., Taniguchi, M., Yoshida, N., Sugiyama, T., and Miura, N. (1997). Essential roles of the winged helix transcription factor MFH-1 in aortic arch patterning and skeletogenesis. *Development* **124**, pp. 4627–38.

Inoue, K., Tanabe, Y., and Lupski, J. R. (1999). Myelin deficiencies in both the central and the peripheral nervous systems associated with a SOX10 mutation. *Ann Neurol* **46**, pp. 313–8.

Inoue, T., Chisaka, O., Matsunami, H., and Takeichi, M. (1997). Cadherin-6 expression transiently delineates specific rhombomeres, other neural tube subdivisions, and neural crest subpopulations in mouse embryos. *Dev Biol* **183**, pp. 183–94.

Inoue, T., Ota, M., Mikoshiba, K., and Aruga, J. (2007). Zic2 and Zic3 synergistically control neurulation and segmentation of paraxial mesoderm in mouse embryo. *Dev Biol* **306**, pp. 669–84.

Ishii, M., Han, J., Yen, H. Y., Sucov, H. M., Chai, Y., and Maxson, R. E., Jr. (2005). Combined deficiencies of Msx1 and Msx2 cause impaired patterning and survival of the cranial neural crest. *Development* **132**, pp. 4937–50.

Ishii, M., Merrill, A. E., Chan, Y. S., Gitelman, I., Rice, D. P., Sucov, H. M., and Maxson, R. E., Jr. (2003). Msx2 and Twist cooperatively control the development of the neural crest-derived skeletogenic mesenchyme of the murine skull vault. *Development* **130**, pp. 6131–42.

Itoh, F., Ishizaka, Y., Tahira, T., Yamamoto, M., Miya, A., Imai, K., Yachi, A., Takai, S., Sugimura, T., and Nagao, M. (1992). Identification and analysis of the ret proto-oncogene promoter region in neuroblastoma cell lines and medullary thyroid carcinomas from MEN2A patients. *Oncogene* **7**, pp. 1201–6.

Itoh, M., Kudoh, T., Dedekian, M., Kim, C. H., and Chitnis, A. B. (2002). A role for iro1 and iro7 in the establishment of an anteroposterior compartment of the ectoderm adjacent to the midbrain–hindbrain boundary. *Development* **129**, pp. 2317–27.

Ittner, L. M., Wurdak, H., Schwerdtfeger, K., Kunz, T., Ille, F., Leveen, P., Hjalt, T. A., Suter, U., Karlsson, S., Hafezi, F., Born, W., and Sommer, L. (2005). Compound developmental eye disorders following inactivation of TGFbeta signaling in neural-crest stem cells. *J Biol* **4**, p. 11.

Ivey, K., Tyson, B., Ukidwe, P., McFadden, D. G., Levi, G., Olson, E. N., Srivastava, D., and Wilkie, T. M. (2003). Galphaq and Galpha11 proteins mediate endothelin-1 signaling in neural crest-derived pharyngeal arch mesenchyme. *Dev Biol* **255**, pp. 230–7.

Ivey, K. N., Sutcliffe, D., Richardson, J., Clyman, R. I., Garcia, J. A., and Srivastava, D. (2008). Transcriptional regulation during development of the ductus arteriosus. *Circ Res* **103**, pp. 388–95.

Iwao, K., Inatani, M., Matsumoto, Y., Ogata-Iwao, M., Takihara, Y., Irie, F., Yamaguchi, Y., Okinami, S., and Tanihara, H. (2009). Heparan sulfate deficiency leads to Peters anomaly in mice by disturbing neural crest TGF-beta2 signaling. *J Clin Invest* **119**, pp. 1997–2008.

Jager, R., Maurer, J., Jacob, A., and Schorle, H. (2004). Cell type-specific conditional regulation of the c-myc proto-oncogene by combining Cre/loxP recombination and tamoxifen-mediated activation. *Genesis* **38**, pp. 145–50.

Jaskoll, T., Zhou, Y. M., Chai, Y., Makarenkova, H. P., Collinson, J. M., West, J. D., Hajihosseini, M. K., Lee, J., and Melnick, M. (2002). Embryonic submandibular gland morphogenesis: stage-specific protein localization of FGFs, BMPs, Pax6 and Pax9 in normal mice and abnormal SMG phenotypes in FgfR2-IIIc(+/Delta), BMP7(-/-) and Pax6(-/-) mice. *Cells Tissues Organs* **170**, pp. 83–98.

Jessen, K. R., and Mirsky, R. (1998). Origin and early development of Schwann cells. *Microsc Res Tech* **41**, pp. 393–402.

Jessen, K. R., and Mirsky, R. (2002). Signals that determine Schwann cell identity. *J Anat* **200**, pp. 367–76.

Ji, M., and Andrisani, O. M. (2005). High-level activation of cyclic AMP signaling attenuates bone morphogenetic protein 2-induced sympathoadrenal lineage development and promotes melanogenesis in neural crest cultures. *Mol Cell Biol* **25**, pp. 5134–45.

Jia, Q., McDill, B. W., Li, S. Z., Deng, C., Chang, C. P., and Chen, F. (2007). Smad signaling in the neural crest regulates cardiac outflow tract remodeling through cell autonomous and non-cell autonomous effects. *Dev Biol* **311**, pp. 172–84.

Jiang, R., Lan, Y., Norton, C. R., Sundberg, J. P., and Gridley, T. (1998a). The Slug gene is not essential for mesoderm or neural crest development in mice. *Dev Biol* **198**, pp. 277–85.

Jiang, R., Norton, C. R., Copeland, N. G., Gilbert, D. J., Jenkins, N. A., and Gridley, T. (1998b). Genomic organization, expression and chromosomal localization of the mouse Slug (Slugh) gene. *Biochim Biophys Acta* **1443**, pp. 251–4.

Jiao, Z., Mollaaghababa, R., Pavan, W. J., Antonellis, A., Green, E. D., and Hornyak, T. J. (2004). Direct interaction of Sox10 with the promoter of murine Dopachrome Tautomerase (Dct) and synergistic activation of Dct expression with Mitf. *Pigment Cell Res* **17**, pp. 352–62.

Jiao, Z., Zhang, Z. G., Hornyak, T. J., Hozeska, A., Zhang, R. L., Wang, Y., Wang, L., Roberts, C., Strickland, F. M., and Chopp, M. (2006). Dopachrome tautomerase (Dct) regulates neural progenitor cell proliferation. *Dev Biol* **296**, pp. 396–408.

Jogi, A., Persson, P., Grynfeld, A., Pahlman, S., and Axelson, H. (2002). Modulation of basic helix–loop–helix transcription complex formation by Id proteins during neuronal differentiation. *J Biol Chem* **277**, pp. 9118–26.

Jogi, A., Vallon-Christersson, J., Holmquist, L., Axelson, H., Borg, A., and Pahlman, S. (2004). Human neuroblastoma cells exposed to hypoxia: induction of genes associated with growth, survival, and aggressive behavior. *Exp Cell Res* **295**, pp. 469–87.

Juergens, K., Rust, B., Pieler, T., and Henningfeld, K. A. (2005). Isolation and comparative expression analysis of the Myc-regulatory proteins Mad1, Mad3, and Mnt during *Xenopus* development. *Dev Dyn* **233**, pp. 1554–9.

Kamholz, J., Awatramani, R., Menichella, D., Jiang, H., Xu, W., and Shy, M. (1999). Regulation of myelin-specific gene expression. Relevance to CMT1. *Ann N Y Acad Sci* **883**, pp. 91–108.

Kanakubo, S., Nomura, T., Yamamura, K., Miyazaki, J., Tamai, M., and Osumi, N. (2006). Abnormal migration and distribution of neural crest cells in Pax6 heterozygous mutant eye, a model for human eye diseases. *Genes Cells* **11**, pp. 919–33.

Kaneko, K. J., Kohn, M. J., Liu, C., and DePamphilis, M. L. (2007). Transcription factor TEAD2 is involved in neural tube closure. *Genesis* **45**, pp. 577–87.

Kao, S. C., Wu, H., Xie, J., Chang, C. P., Ranish, J. A., Graef, I. A., and Crabtree, G. R. (2009). Calcineurin/NFAT signaling is required for neuregulin-regulated Schwann cell differentiation. *Science* **323**, pp. 651–4.

Kapur, R. P. (1999). Early death of neural crest cells is responsible for total enteric aganglionosis in Sox10(Dom)/Sox10(Dom) mouse embryos. *Pediatr Dev Pathol* **2**, pp. 559–69.

Karafiat, V., Dvorakova, M., Krejci, E., Kralova, J., Pajer, P., Snajdr, P., Mandikova, S., Bartunek, P., Grim, M., and Dvorak, M. (2005). Transcription factor c-Myb is involved in the regulation of the epithelial–mesenchymal transition in the avian neural crest. *Cell Mol Life Sci* **62**, pp. 2516–25.

Karaman, A., and Aliagaoglu, C. (2006). Waardenburg syndrome type 1. *Dermatol Online J* **12**, p. 21.

Karavanov, A. A., Saint-Jeannet, J. P., Karavanova, I., Taira, M., and Dawid, I. B. (1996). The LIM homeodomain protein Lim-1 is widely expressed in neural, neural crest and mesoderm derivatives in vertebrate development. *Int J Dev Biol* **40**, pp. 453–61.

Karg, H., Burger, E. H., Lyaruu, D. M., Bronckers, A. L., and Woltgens, J. H. (1997). Spatiotemporal expression of the homeobox gene S8 during mouse tooth development. *Arch Oral Biol* **42**, pp. 625–31.

Kawakami, A., Kimura-Kawakami, M., Nomura, T., and Fujisawa, H. (1997). Distributions of PAX6 and PAX7 proteins suggest their involvement in both early and late phases of chick brain development. *Mech Dev* **66**, pp. 119–30.

Kazama, H., Kodera, T., Shimizu, S., Mizoguchi, H., and Morishita, K. (1999). Ecotropic viral integration site-1 is activated during, and is sufficient for, neuroectodermal P19 cell differentiation. *Cell Growth Differ* **10**, pp. 565–73.

Keller, M. J., and Chitnis, A. B. (2007). Insights into the evolutionary history of the vertebrate zic3 locus from a teleost-specific zic6 gene in the zebrafish, Danio rerio. *Dev Genes Evol* **217**, pp. 541–7.

Kelsh, R. N. (2006). Sorting out Sox10 functions in neural crest development. *Bioessays* **28**, pp. 788–98.

Kerney, R., Gross, J. B., and Hanken, J. (2007). Runx2 is essential for larval hyobranchial cartilage formation in *Xenopus laevis*. *Dev Dyn* **236**, pp. 1650–62.

Khadka, D., Luo, T., and Sargent, T. D. (2006). Msx1 and Msx2 have shared essential functions in neural crest but may be dispensable in epidermis and axis formation in *Xenopus*. *Int J Dev Biol* **50**, pp. 499–502.

Kim, J., Lo, L., Dormand, E., and Anderson, D. J. (2003). SOX10 maintains multipotency and inhibits neuronal differentiation of neural crest stem cells. *Neuron* **38**, pp. 17–31.

Kioussi, C., Gross, M. K., and Gruss, P. (1995). Pax3: a paired domain gene as a regulator in PNS myelination. *Neuron* **15**, pp. 553–62.

Kirby, M. L. (1990). Alteration of cardiogenesis after neural crest ablation. *Ann N Y Acad Sci* **588**, pp. 289–95.

Kirby, M. L., Hunt, P., Wallis, K., and Thorogood, P. (1997). Abnormal patterning of the aortic arch arteries does not evoke cardiac malformations. *Dev Dyn* **208**, pp. 34–47.

Kist, R., Greally, E., and Peters, H. (2007). Derivation of a mouse model for conditional inactivation of Pax9. *Genesis* **45**, pp. 460–4.

Kitaguchi, T., Nakata, K., Nagai, T., Aruga, J., and Mikoshiba, K. (2001). *Xenopus* Polycomblike 2 (XPcl2) controls anterior to posterior patterning of the neural tissue. *Dev Genes Evol* **211**, pp. 309–14.

Kitahashi, M., Sato, Y., Fujimura, L., Ozeki, C., Arima, M., Sakamoto, A., Yamamoto, S., Tokuhisa, T., and Hatano, M. (2007). Identification of the consensus DNA sequence for Nczf binding. *DNA Cell Biol* **26**, pp. 395–401.

Knezevich, S. R., Hendson, G., Mathers, J. A., Carpenter, B., Lopez-Terrada, D., Brown, K. L., and Sorensen, P. H. (1998). Absence of detectable EWS/FLI1 expression after therapy-induced neural differentiation in Ewing sarcoma. *Hum Pathol* **29**, pp. 289–94.

Knight, R. D., Javidan, Y., Nelson, S., Zhang, T., and Schilling, T. (2004). Skeletal and pigment cell defects in the lockjaw mutant reveal multiple roles for zebrafish tfap2a in neural crest development. *Dev Dyn* **229**, pp. 87–98.

Knight, R. D., Javidan, Y., Zhang, T., Nelson, S., and Schilling, T. F. (2005). AP2-dependent signals from the ectoderm regulate craniofacial development in the zebrafish embryo. *Development* **132**, pp. 3127–38.

Knight, R. D., Nair, S., Nelson, S. S., Afshar, A., Javidan, Y., Geisler, R., Rauch, G. J., and Schilling, T. F. (2003). lockjaw encodes a zebrafish tfap2a required for early neural crest development. *Development* **130**, pp. 5755–68.

Kobayashi, M., Hjerling-Leffler, J., and Ernfors, P. (2006). Increased progenitor proliferation and apoptotic cell death in the sensory lineage of mice overexpressing N-myc. *Cell Tissue Res* **323**, pp. 81–90.

Koblar, S. A., Murphy, M., Barrett, G. L., Underhill, A., Gros, P., and Bartlett, P. F. (1999). Pax-3 regulates neurogenesis in neural crest-derived precursor cells. *J Neurosci Res* **56**, pp. 518–30.

Kohlbecker, A., Lee, A. E., and Schorle, H. (2002). Exencephaly in a subset of animals heterozygous for AP-2alpha mutation. *Teratology* **65**, pp. 213–8.

Koppen, A., Ait-Aissa, R., Hopman, S., Koster, J., Haneveld, F., Versteeg, R., and Valentijn, L. J. (2007). Dickkopf-1 is down-regulated by MYCN and inhibits neuroblastoma cell proliferation. *Cancer Lett* **256**, pp. 218–28.

Kos, R., Reedy, M. V., Johnson, R. L., and Erickson, C. A. (2001). The winged-helix transcription factor FoxD3 is important for establishing the neural crest lineage and repressing melanogenesis in avian embryos. *Development* **128**, pp. 1467–79.

Koster, M., Dillinger, K., and Knochel, W. (1998). Expression pattern of the winged helix factor XFD-11 during *Xenopus* embryogenesis. *Mech Dev* **76**, pp. 169–73.

Koster, M., Dillinger, K., and Knochel, W. (1999). Genomic structure and embryonic expression of the *Xenopus* winged helix factors XFD-13/13'. *Mech Dev* **88**, pp. 89–93.

Kraus, P., and Lufkin, T. (2006). Dlx homeobox gene control of mammalian limb and craniofacial development. *Am J Med Genet A* **140**, pp. 1366–74.

Kuhlbrodt, K., Herbarth, B., Sock, E., Hermans-Borgmeyer, I., and Wegner, M. (1998a). Sox10, a novel transcriptional modulator in glial cells. *J Neurosci* **18**, pp. 237–50.

Kuhlbrodt, K., Schmidt, C., Sock, E., Pingault, V., Bondurand, N., Goossens, M., and Wegner, M. (1998b). Functional analysis of Sox10 mutations found in human Waardenburg–Hirschsprung patients. *J Biol Chem* **273**, pp. 23033–8.

Kumar, S. D., Dheen, S. T., and Tay, S. S. (2007). Maternal diabetes induces congenital heart defects in mice by altering the expression of genes involved in cardiovascular development. *Cardiovasc Diabetol* **6**, p. 34.

Kumasaka, M., Sato, H., Sato, S., Yajima, I., and Yamamoto, H. (2004). Isolation and developmental expression of Mitf in *Xenopus laevis*. *Dev Dyn* **230**, pp. 107–13.

Kume, T., Jiang, H., Topczewska, J. M., and Hogan, B. L. (2001). The murine winged helix transcription factors, Foxc1 and Foxc2, are both required for cardiovascular development and somitogenesis. *Genes Dev* **15**, pp. 2470–82.

Kuo, J. S., Patel, M., Gamse, J., Merzdorf, C., Liu, X., Apekin, V., and Sive, H. (1998). Opl: a zinc finger protein that regulates neural determination and patterning in *Xenopus*. *Development* **125**, pp. 2867–82.

Kuphal, S., Palm, H. G., Poser, I., and Bosserhoff, A. K. (2005). Snail-regulated genes in malignant melanoma. *Melanoma Res* **15**, pp. 305–13.

Kurata, T., and Ueno, N. (2003). *Xenopus* Nbx, a novel NK-1 related gene essential for neural crest formation. *Dev Biol* **257**, pp. 30–40.

Kuriyama, S., and Mayor, R. (2008). Molecular analysis of neural crest migration. *Philos Trans R Soc Lond B Biol Sci* **363**, pp. 1349–62.

Kwang, S. J., Brugger, S. M., Lazik, A., Merrill, A. E., Wu, L. Y., Liu, Y. H., Ishii, M., Sangiorgi, F. O., Rauchman, M., Sucov, H. M., Maas, R. L., and Maxson, R. E., Jr. (2002). Msx2 is an immediate downstream effector of Pax3 in the development of the murine cardiac neural crest. *Development* **129**, pp. 527–38.

LaBonne, C., and Bronner-Fraser, M. (1998). Neural crest induction in *Xenopus*: evidence for a two-signal model. *Development* **125**, pp. 2403–14.

LaBonne, C., and Bronner-Fraser, M. (2000). Snail-related transcriptional repressors are required in *Xenopus* for both the induction of the neural crest and its subsequent migration. *Dev Biol* **221**, pp. 195–205.

Labosky, P. A., and Kaestner, K. H. (1998). The winged helix transcription factor Hfh2 is expressed in neural crest and spinal cord during mouse development. *Mech Dev* **76**, pp. 185–90.

Lacosta, A. M., Canudas, J., Gonzalez, C., Muniesa, P., Sarasa, M., and Dominguez, L. (2007). Pax7 identifies neural crest, chromatophore lineages and pigment stem cells during zebrafish development. *Int J Dev Biol* **51**, pp. 327–31.

Lacosta, A. M., Muniesa, P., Ruberte, J., Sarasa, M., and Dominguez, L. (2005). Novel expression patterns of Pax3/Pax7 in early trunk neural crest and its melanocyte and non-melanocyte lineages in amniote embryos. *Pigment Cell Res* **18**, pp. 243–51.

Lakkis, M. M., Golden, J. A., O'Shea, K. S., and Epstein, J. A. (1999). Neurofibromin deficiency in mice causes exencephaly and is a modifier for Splotch neural tube defects. *Dev Biol* **212**, pp. 80–92.

Lakshmanan, G., Lieuw, K. H., Lim, K. C., Gu, Y., Grosveld, F., Engel, J. D., and Karis, A. (1999). Localization of distant urogenital system-, central nervous system-, and endocardium-specific transcriptional regulatory elements in the GATA-3 locus. *Mol Cell Biol* **19**, pp. 1558–68.

Lalwani, A. K., Attaie, A., Randolph, F. T., Deshmukh, D., Wang, C., Mhatre, A., and Wilcox, E. (1998). Point mutation in the MITF gene causing Waardenburg syndrome type II in a three-generation Indian family. *Am J Med Genet* **80**, pp. 406–9.

Lan, Y., and Jiang, R. (2009). Sonic hedgehog signaling regulates reciprocal epithelial–mesenchymal interactions controlling palatal outgrowth. *Development* **136**, pp. 1387–96.

Landgren, H., and Carlsson, P. (2004). FoxJ3, a novel mammalian forkhead gene expressed in neuroectoderm, neural crest, and myotome. *Dev Dyn* **231**, pp. 396–401.

Lang, D., Brown, C. B., Milewski, R., Jiang, Y. Q., Lu, M. M., and Epstein, J. A. (2003). Distinct enhancers regulate neural expression of Pax7. *Genomics* **82**, pp. 553–60.

Lang, D., Chen, F., Milewski, R., Li, J., Lu, M. M., and Epstein, J. A. (2000). Pax3 is required for enteric ganglia formation and functions with Sox10 to modulate expression of c-ret. *J Clin Invest* **106**, pp. 963–71.

Lang, D., and Epstein, J. A. (2003). Sox10 and Pax3 physically interact to mediate activation of a conserved c-RET enhancer. *Hum Mol Genet* **12**, pp. 937–45.

Lang, D., Lu, M. M., Huang, L., Engleka, K. A., Zhang, M., Chu, E. Y., Lipner, S., Skoultchi, A., Millar, S. E., and Epstein, J. A. (2005). Pax3 functions at a nodal point in melanocyte stem cell differentiation. *Nature* **433**, pp. 884–7.

Langer, E. M., Feng, Y., Zhaoyuan, H., Rauscher, F. J., 3rd, Kroll, K. L., and Longmore, G. D. (2008). Ajuba LIM proteins are snail/slug corepressors required for neural crest development in *Xenopus*. *Dev Cell* **14**, pp. 424–36.

Lanigan, T. M., DeRaad, S. K., and Russo, A. F. (1998). Requirement of the MASH-1 transcription factor for neuroendocrine differentiation of thyroid C cells. *J Neurobiol* **34**, pp. 126–34.

Larue, L., and Delmas, V. (2006). The WNT/Beta-catenin pathway in melanoma. *Front Biosci* **11**, pp. 733–42.

Larue, L., Kumasaka, M., and Goding, C. R. (2003). Beta-catenin in the melanocyte lineage. *Pigment Cell Res* **16**, pp. 312–7.

Lasorella, A., and Iavarone, A. (2006). The protein ENH is a cytoplasmic sequestration factor for Id2 in normal and tumor cells from the nervous system. *Proc Natl Acad Sci U S A* **103**, pp. 4976–81.

Lasorella, A., Noseda, M., Beyna, M., Yokota, Y., and Iavarone, A. (2000). Id2 is a retinoblastoma protein target and mediates signalling by Myc oncoproteins. *Nature* **407**, pp. 592–8.

Lawson, A., Colas, J. F., and Schoenwolf, G. C. (2000). Ectodermal markers delineate the neural fold interface during avian neurulation. *Anat Rec* **260**, pp. 106–9.

Lawson, A., Schoenwolf, G. C., England, M. A., Addai, F. K., and Ahima, R. S. (1999). Programmed cell death and the morphogenesis of the hindbrain roof plate in the chick embryo. *Anat Embryol (Berl)* **200**, pp. 509–19.

Le Douarin, N. M. (1980). The ontogeny of the neural crest in avian embryo chimaeras. *Nature* **286**, pp. 663–9.

Le Douarin, N. M., Calloni, G. W., and Dupin, E. (2008). The stem cells of the neural crest. *Cell Cycle* **7**, pp. 1013–9.

Lecaudey, V., Thisse, C., Thisse, B., and Schneider-Maunoury, S. (2001). Sequence and expression pattern of ziro7, a novel, divergent zebrafish iroquois homeobox gene. *Mech Dev* **109**, pp. 383–8.

Lee, J. A., Anholt, R. R., and Cole, G. J. (2008a). Olfactomedin-2 mediates development of the anterior central nervous system and head structures in zebrafish. *Mech Dev* **125**, pp. 167–81.

Lee, K. E., Nam, S., Cho, E. A., Seong, I., Limb, J. K., Lee, S., and Kim, J. (2008b). Identification of direct regulatory targets of the transcription factor Sox10 based on function and conservation. *BMC Genomics* **9**, p. 408.

Lee, S. K., Jurata, L. W., Nowak, R., Lettieri, K., Kenny, D. A., Pfaff, S. L., and Gill, G. N. (2005a). The LIM domain-only protein LMO4 is required for neural tube closure. *Mol Cell Neurosci* **28**, pp. 205–14.

Lee, S. Y., Lee, H. S., Moon, J. S., Kim, J. I., Park, J. B., Lee, J. Y., Park, M. J., and Kim, J. (2004a). Transcriptional regulation of Zic3 by heterodimeric AP-1(c-Jun/c-Fos) during *Xenopus* development. *Exp Mol Med* **36**, pp. 468–75.

Lee, V. M., Bronner-Fraser, M., and Baker, C. V. (2005b). Restricted response of mesencephalic neural crest to sympathetic differentiation signals in the trunk. *Dev Biol* **278**, pp. 175–92.

Lee, Y. H., Aoki, Y., Hong, C. S., Saint-Germain, N., Credidio, C., and Saint-Jeannet, J. P. (2004b). Early requirement of the transcriptional activator Sox9 for neural crest specification in *Xenopus*. *Dev Biol* **275**, pp. 93–103.

Lee, Y. H., and Saint-Jeannet, J. P. (2009). Characterization of molecular markers to assess cardiac cushions formation in *Xenopus*. *Dev Dyn* **238**, pp. 3257–65.

Lefebvre, V., Li, P., and de Crombrugghe, B. (1998). A new long form of Sox5 (L-Sox5), Sox6 and Sox9 are coexpressed in chondrogenesis and cooperatively activate the type II collagen gene. *Embo J* **17**, pp. 5718–33.

Lepore, J. J., Mericko, P. A., Cheng, L., Lu, M. M., Morrisey, E. E., and Parmacek, M. S. (2006). GATA-6 regulates semaphorin 3C and is required in cardiac neural crest for cardiovascular morphogenesis. *J Clin Invest* **116**, pp. 929–39.

Li, F., Luo, Z., Huang, W., Lu, Q., Wilcox, C. S., Jose, P. A., and Chen, S. (2007). Response gene to complement 32, a novel regulator for transforming growth factor-beta-induced smooth muscle differentiation of neural crest cells. *J Biol Chem* **282**, pp. 10133–7.

Li, J., Chen, F., and Epstein, J. A. (2000). Neural crest expression of Cre recombinase directed by the proximal Pax3 promoter in transgenic mice. *Genesis* **26**, pp. 162–4.

Li, J., Liu, K. C., Jin, F., Lu, M. M., and Epstein, J. A. (1999). Transgenic rescue of congenital heart disease and spina bifida in Splotch mice. *Development* **126**, pp. 2495–503.

Li, J., Molkentin, J. D., and Colbert, M. C. (2001). Retinoic acid inhibits cardiac neural crest migration by blocking c-Jun N-terminal kinase activation. *Dev Biol* **232**, pp. 351–61.

Li, J., Zhu, X., Chen, M., Cheng, L., Zhou, D., Lu, M. M., Du, K., Epstein, J. A., and Parmacek, M. S. (2005). Myocardin-related transcription factor B is required in cardiac neural crest for smooth muscle differentiation and cardiovascular development. *Proc Natl Acad Sci U S A* **102**, pp. 8916–21.

Li, M., Zhao, C., Wang, Y., Zhao, Z., and Meng, A. (2002). Zebrafish sox9b is an early neural crest marker. *Dev Genes Evol* **212**, pp. 203–6.

Li, W., and Cornell, R. A. (2007). Redundant activities of Tfap2a and Tfap2c are required for neural crest induction and development of other non-neural ectoderm derivatives in zebrafish embryos. *Dev Biol* **304**, pp. 338–54.

Liang, X., Sun, Y., Ye, M., Scimia, M. C., Cheng, H., Martin, J., Wang, G., Rearden, A., Wu, C., Peterson, K. L., Powell, H. C., Evans, S. M., and Chen, J. (2009). Targeted ablation of PINCH1 and PINCH2 from murine myocardium results in dilated cardiomyopathy and early postnatal lethality. *Circulation* **120**, pp. 568–76.

Lichty, B. D., Ackland-Snow, J., Noble, L., Kamel-Reid, S., and Dube, I. D. (1995). Dysregulation of HOX11 by chromosome translocations in T-cell acute lymphoblastic leukemia: a paradigm for homeobox gene involvement in human cancer. *Leuk Lymphoma* **16**, pp. 209–15.

Light, W., Vernon, A. E., Lasorella, A., Iavarone, A., and LaBonne, C. (2005). *Xenopus* Id3 is required downstream of Myc for the formation of multipotent neural crest progenitor cells. *Development* **132**, pp. 1831–41.

Lim, K. C., Lakshmanan, G., Crawford, S. E., Gu, Y., Grosveld, F., and Engel, J. D. (2000). Gata3 loss leads to embryonic lethality due to noradrenaline deficiency of the sympathetic nervous system. *Nat Genet* **25**, pp. 209–12.

Lines, M. A., Kozlowski, K., and Walter, M. A. (2002). Molecular genetics of Axenfeld–Rieger malformations. *Hum Mol Genet* **11**, pp. 1177–84.

Linker, C., Bronner-Fraser, M., and Mayor, R. (2000). Relationship between gene expression domains of Xsnail, Xslug, and Xtwist and cell movement in the prospective neural crest of *Xenopus*. *Dev Biol* **224**, pp. 215–25.

Lister, J. A. (2002). Development of pigment cells in the zebrafish embryo. *Microsc Res Tech* **58**, pp. 435–41.

Lister, J. A., Close, J., and Raible, D. W. (2001). Duplicate mitf genes in zebrafish: complementary expression and conservation of melanogenic potential. *Dev Biol* **237**, pp. 333–44.

Lister, J. A., Cooper, C., Nguyen, K., Modrell, M., Grant, K., and Raible, D. W. (2006). Zebrafish Foxd3 is required for development of a subset of neural crest derivatives. *Dev Biol* **290**, pp. 92–104.

Lister, J. A., Robertson, C. P., Lepage, T., Johnson, S. L., and Raible, D. W. (1999). nacre encodes a zebrafish microphthalmia-related protein that regulates neural-crest-derived pigment cell fate. *Development* **126**, pp. 3757–67.

Liu, H., Margiotta, J. F., and Howard, M. J. (2005). BMP4 supports noradrenergic differentiation by a PKA-dependent mechanism. *Dev Biol* **286**, pp. 521–36.

Liu, Q., Melnikova, I. N., Hu, M., and Gardner, P. D. (1999). Cell type-specific activation of neuronal nicotinic acetylcholine receptor subunit genes by Sox10. *J Neurosci* **19**, pp. 9747–55.

Lo, L., Dormand, E., Greenwood, A., and Anderson, D. J. (2002). Comparison of the generic neuronal differentiation and neuron subtype specification functions of mammalian achaete-scute and atonal homologs in cultured neural progenitor cells. *Development* **129**, pp. 1553–67.

Lo, L., Guillemot, F., Joyner, A. L., and Anderson, D. J. (1994). MASH-1: a marker and a mutation for mammalian neural crest development. *Perspect Dev Neurobiol* **2**, pp. 191–201.

Lo, L., Sommer, L., and Anderson, D. J. (1997). MASH1 maintains competence for BMP2-induced neuronal differentiation in post-migratory neural crest cells. *Curr Biol* **7**, pp. 440–50.

Lo, L., Tiveron, M. C., and Anderson, D. J. (1998). MASH1 activates expression of the paired homeodomain transcription factor Phox2a, and couples pan-neuronal and subtype-specific components of autonomic neuronal identity. *Development* **125**, pp. 609–20.

Lo, L. C., Johnson, J. E., Wuenschell, C. W., Saito, T., and Anderson, D. J. (1991). Mammalian achaete-scute homolog 1 is transiently expressed by spatially restricted subsets of early neuroepithelial and neural crest cells. *Genes Dev* **5**, pp. 1524–37.

Lobsiger, C. S., Smith, P. M., Buchstaller, J., Schweitzer, B., Franklin, R. J., Suter, U., and Taylor, V. (2001). SpL201: a conditionally immortalized Schwann cell precursor line that generates myelin. *Glia* **36**, pp. 31–47.

Locascio, A., Vega, S., de Frutos, C. A., Manzanares, M., and Nieto, M. A. (2002). Biological potential of a functional human SNAIL retrogene. *J Biol Chem* **277**, pp. 38803–9.

Logan, C., Wingate, R. J., McKay, I. J., and Lumsden, A. (1998). Tlx-1 and Tlx-3 homeobox gene expression in cranial sensory ganglia and hindbrain of the chick embryo: markers of patterned connectivity. *J Neurosci* **18**, pp. 5389–402.

Lopes, S. S., Yang, X., Muller, J., Carney, T. J., McAdow, A. R., Rauch, G. J., Jacoby, A. S., Hurst, L. D., Delfino-Machin, M., Haffter, P., Geisler, R., Johnson, S. L., Ward, A., and Kelsh, R. N. (2008). Leukocyte tyrosine kinase functions in pigment cell development. *PLoS Genet* **4**, p. e1000026.

Lopez-Schier, H., and Hudspeth, A. J. (2005). Supernumerary neuromasts in the posterior lateral line of zebrafish lacking peripheral glia. *Proc Natl Acad Sci U S A* **102**, pp. 1496–501.

Lucas, M. E., Muller, F., Rudiger, R., Henion, P. D., and Rohrer, H. (2006). The bHLH transcription factor hand2 is essential for noradrenergic differentiation of sympathetic neurons. *Development* **133**, pp. 4015–24.

Lui, V. C., Cheng, W. W., Leon, T. Y., Lau, D. K., Garcia-Barcelo, M. M., Miao, X. P., Kam, M. K., So, M. T., Chen, Y., Wall, N. A., Sham, M. H., and Tam, P. K. (2008). Perturbation of hoxb5 signaling in vagal neural crests down-regulates ret leading to intestinal hypoganglionosis in mice. *Gastroenterology* **134**, pp. 1104–15.

Luo, T., Lee, Y. H., Saint-Jeannet, J. P., and Sargent, T. D. (2003). Induction of neural crest in *Xenopus* by transcription factor AP2alpha. *Proc Natl Acad Sci U S A* **100**, pp. 532–7.

Luo, T., Matsuo-Takasaki, M., Thomas, M. L., Weeks, D. L., and Sargent, T. D. (2002). Transcription factor AP-2 is an essential and direct regulator of epidermal development in *Xenopus*. *Dev Biol* **245**, pp. 136–44.

Luo, T., Xu, Y., Hoffman, T. L., Zhang, T., Schilling, T., and Sargent, T. D. (2007). Inca: a novel p21-activated kinase-associated protein required for cranial neural crest development. *Development* **134**, pp. 1279–89.

Luo, T., Zhang, Y., Khadka, D., Rangarajan, J., Cho, K. W., and Sargent, T. D. (2005). Regulatory targets for transcription factor AP2 in *Xenopus* embryos. *Dev Growth Differ* **47**, pp. 403–13.

Lyons, G. E., Micales, B. K., Schwarz, J., Martin, J. F., and Olson, E. N. (1995). Expression of mef2 genes in the mouse central nervous system suggests a role in neuronal maturation. *J Neurosci* **15**, pp. 5727–38.

MacArthur, L. (1996). AP-1-related proteins bind to the enkephalin CRE-2 element in adrenal chromaffin cells. *J Neurochem* **67**, pp. 2256–64.

MacDonald, S. T., Bamforth, S. D., Chen, C. M., Farthing, C. R., Franklyn, A., Broadbent, C., Schneider, J. E., Saga, Y., Lewandoski, M., and Bhattacharya, S. (2008). Epiblastic Cited2 deficiency results in cardiac phenotypic heterogeneity and provides a mechanism for haploinsufficiency. *Cardiovasc Res* **79**, pp. 448–57.

Maconochie, M., Krishnamurthy, R., Nonchev, S., Meier, P., Manzanares, M., Mitchell, P. J., and Krumlauf, R. (1999). Regulation of Hoxa2 in cranial neural crest cells involves members of the AP-2 family. *Development* **126**, pp. 1483–94.

Maeda, R., Mood, K., Jones, T. L., Aruga, J., Buchberg, A. M., and Daar, I. O. (2001). Xmeis1, a protooncogene involved in specifying neural crest cell fate in *Xenopus* embryos. *Oncogene* **20**, pp. 1329–42.

Mager, A. M., Grapin-Botton, A., Ladjali, K., Meyer, D., Wolff, C. M., Stiegler, P., Bonnin, M. A., and Remy, P. (1998). The avian fli gene is specifically expressed during embryogenesis in a subset of neural crest cells giving rise to mesenchyme. *Int J Dev Biol* **42**, pp. 561–72.

Magnaghi, P., Roberts, C., Lorain, S., Lipinski, M., and Scambler, P. J. (1998). HIRA, a mammalian homologue of *Saccharomyces cerevisiae* transcriptional co-repressors, interacts with Pax3. *Nat Genet* **20**, pp. 74–7.

Maka, M., Stolt, C. C., and Wegner, M. (2005). Identification of Sox8 as a modifier gene in a mouse model of Hirschsprung disease reveals underlying molecular defect. *Dev Biol* **277**, pp. 155–69.

Makarenkova, H. P., Ito, M., Govindarajan, V., Faber, S. C., Sun, L., McMahon, G., Overbeek,

P. A., and Lang, R. A. (2000). FGF10 is an inducer and Pax6 a competence factor for lacrimal gland development. *Development* **127**, pp. 2563–72.

Mallo, M. (2001). Formation of the middle ear: recent progress on the developmental and molecular mechanisms. *Dev Biol* **231**, pp. 410–9.

Mani, A., Radhakrishnan, J., Farhi, A., Carew, K. S., Warnes, C. A., Nelson-Williams, C., Day, R. W., Pober, B., State, M. W., and Lifton, R. P. (2005). Syndromic patent ductus arteriosus: evidence for haploinsufficient TFAP2B mutations and identification of a linked sleep disorder. *Proc Natl Acad Sci U S A* **102**, pp. 2975–9.

Manley, N. R., and Capecchi, M. R. (1997). Hox group 3 paralogous genes act synergistically in the formation of somitic and neural crest-derived structures. *Dev Biol* **192**, pp. 274–88.

Mansouri, A., Pla, P., Larue, L., and Gruss, P. (2001). Pax3 acts cell autonomously in the neural tube and somites by controlling cell surface properties. *Development* **128**, pp. 1995–2005.

Mansouri, A., Stoykova, A., Torres, M., and Gruss, P. (1996). Dysgenesis of cephalic neural crest derivatives in Pax7-/- mutant mice. *Development* **122**, pp. 831–8.

Marazzi, G., Wang, Y., and Sassoon, D. (1997). Msx2 is a transcriptional regulator in the BMP4-mediated programmed cell death pathway. *Dev Biol* **186**, pp. 127–38.

Marin, F., and Charnay, P. (2000). Hindbrain patterning: FGFs regulate Krox20 and mafB/kr expression in the otic/preotic region. *Development* **127**, pp. 4925–35.

Mark, M., Ghyselinck, N. B., Kastner, P., Dupe, V., Wendling, O., Krezel, W., Mascrez, B., and Chambon, P. (1998). Mesectoderm is a major target of retinoic acid action. *Eur J Oral Sci* **106 Suppl 1**, pp. 24–31.

Marmigere, F., Montelius, A., Wegner, M., Groner, Y., Reichardt, L. F., and Ernfors, P. (2006). The Runx1/AML1 transcription factor selectively regulates development and survival of TrkA nociceptive sensory neurons. *Nat Neurosci* **9**, pp. 180–7.

Maro, G. S., Vermeren, M., Voiculescu, O., Melton, L., Cohen, J., Charnay, P., and Topilko, P. (2004). Neural crest boundary cap cells constitute a source of neuronal and glial cells of the PNS. *Nat Neurosci* **7**, pp. 930–8.

Maroulakou, I. G., Papas, T. S., and Green, J. E. (1994). Differential expression of ets-1 and ets-2 proto-oncogenes during murine embryogenesis. *Oncogene* **9**, pp. 1551–65.

Marshall, H., Nonchev, S., Sham, M. H., Muchamore, I., Lumsden, A., and Krumlauf, R. (1992). Retinoic acid alters hindbrain Hox code and induces transformation of rhombomeres 2/3 into a 4/5 identity. *Nature* **360**, pp. 737–41.

Martin, J. F., Bradley, A., and Olson, E. N. (1995). The paired-like homeo box gene MHox is required for early events of skeletogenesis in multiple lineages. *Genes Dev* **9**, pp. 1237–49.

Martinsen, B. J., Neumann, A. N., Frasier, A. J., Baker, C. V., Krull, C. E., and Lohr, J. L. (2006). PINCH-1 expression during early avian embryogenesis: implications for neural crest and heart development. *Dev Dyn* **235**, pp. 152–62.

Maschhoff, K. L., Anziano, P. Q., Ward, P., and Baldwin, H. S. (2003). Conservation of Sox4 gene structure and expression during chicken embryogenesis. *Gene* **320**, pp. 23–30.

Matera, I., Watkins-Chow, D. E., Loftus, S. K., Hou, L., Incao, A., Silver, D. L., Rivas, C., Elliott, E. C., Baxter, L. L., and Pavan, W. J. (2008). A sensitized mutagenesis screen identifies Gli3 as a modifier of Sox10 neurocristopathy. *Hum Mol Genet* **17**, pp. 2118–31.

Matheny, C., DiStefano, P. S., and Milbrandt, J. (1992). Differential activation of NGF receptor and early response genes in neural crest-derived cells. *Brain Res Mol Brain Res* **13**, pp. 75–81.

Matsuo, T., Osumi-Yamashita, N., Noji, S., Ohuchi, H., Koyama, E., Myokai, F., Matsuo, N., Taniguchi, S., Doi, H., Iseki, S., and et al. (1993). A mutation in the Pax-6 gene in rat small eye is associated with impaired migration of midbrain crest cells. *Nat Genet* **3**, pp. 299–304.

Matsushima, Y., Shinkai, Y., Kobayashi, Y., Sakamoto, M., Kunieda, T., and Tachibana, M. (2002). A mouse model of Waardenburg syndrome type 4 with a new spontaneous mutation of the endothelin-B receptor gene. *Mamm Genome* **13**, pp. 30–5.

Matt, N., Ghyselinck, N. B., Pellerin, I., and Dupe, V. (2008). Impairing retinoic acid signalling in the neural crest cells is sufficient to alter entire eye morphogenesis. *Dev Biol* **320**, pp. 140–8.

Matt, N., Ghyselinck, N. B., Wendling, O., Chambon, P., and Mark, M. (2003). Retinoic acid-induced developmental defects are mediated by RARbeta/RXR heterodimers in the pharyngeal endoderm. *Development* **130**, pp. 2083–93.

Maurer, J., Fuchs, S., Jager, R., Kurz, B., Sommer, L., and Schorle, H. (2007). Establishment and controlled differentiation of neural crest stem cell lines using conditional transgenesis. *Differentiation* **75**, pp. 580–91.

Mayanil, C. S., George, D., Freilich, L., Miljan, E. J., Mania-Farnell, B., McLone, D. G., and Bremer, E. G. (2001). Microarray analysis detects novel Pax3 downstream target genes. *J Biol Chem* **276**, pp. 49299–309.

Mayor, R., and Aybar, M. J. (2001). Induction and development of neural crest in *Xenopus laevis*. *Cell Tissue Res* **305**, pp. 203–9.

Mayor, R., Guerrero, N., Young, R. M., Gomez-Skarmeta, J. L., and Cuellar, C. (2000). A novel function for the Xslug gene: control of dorsal mesendoderm development by repressing BMP-4. *Mech Dev* **97**, pp. 47–56.

Mayor, R., Morgan, R., and Sargent, M. G. (1995). Induction of the prospective neural crest of *Xenopus*. *Development* **121**, pp. 767–77.

McFadden, D. G., Charite, J., Richardson, J. A., Srivastava, D., Firulli, A. B., and Olson, E. N. (2000). A GATA-dependent right ventricular enhancer controls dHAND transcription in the developing heart. *Development* **127**, pp. 5331–41.

McGonnell, I. M., McKay, I. J., and Graham, A. (2001). A population of caudally migrating cranial neural crest cells: functional and evolutionary implications. *Dev Biol* **236**, pp. 354–63.

McGuinness, T., Porteus, M. H., Smiga, S., Bulfone, A., Kingsley, C., Qiu, M., Liu, J. K., Long, J. E., Xu, D., and Rubenstein, J. L. (1996). Sequence, organization, and transcription of the Dlx-1 and Dlx-2 locus. *Genomics* **35**, pp. 473–85.

McKeown, S. J., Lee, V. M., Bronner-Fraser, M., Newgreen, D. F., and Farlie, P. G. (2005). Sox10 overexpression induces neural crest-like cells from all dorsoventral levels of the neural tube but inhibits differentiation. *Dev Dyn* **233**, pp. 430–44.

McNulty, C. L., Peres, J. N., Bardine, N., van den Akker, W. M., and Durston, A. J. (2005). Knockdown of the complete Hox paralogous group 1 leads to dramatic hindbrain and neural crest defects. *Development* **132**, pp. 2861–71.

McPherson, C. E., Varley, J. E., and Maxwell, G. D. (2000). Expression and regulation of type I BMP receptors during early avian sympathetic ganglion development. *Dev Biol* **221**, pp. 220–32.

Mead, P. E., Parganas, E., Ohtsuka, S., Morishita, K., Gamer, L., Kuliyev, E., Wright, C. V., and Ihle, J. N. (2005). Evi-1 expression in *Xenopus. Gene Expr Patterns* **5**, pp. 601–8.

Menegola, E., Broccia, M. L., Di Renzo, F., Massa, V., and Giavini, E. (2004). Relationship between hindbrain segmentation, neural crest cell migration and branchial arch abnormalities in rat embryos exposed to fluconazole and retinoic acid in vitro. *Reprod Toxicol* **18**, pp. 121–30.

Menegola, E., Broccia, M. L., Di Renzo, F., Massa, V., and Giavini, E. (2005). Study on the common teratogenic pathway elicited by the fungicides triazole-derivatives. *Toxicol In Vitro* **19**, pp. 737–48.

Merlo, G. R., Zerega, B., Paleari, L., Trombino, S., Mantero, S., and Levi, G. (2000). Multiple functions of Dlx genes. *Int J Dev Biol* **44**, pp. 619–26.

Merrill, A. E., Bochukova, E. G., Brugger, S. M., Ishii, M., Pilz, D. T., Wall, S. A., Lyons, K. M., Wilkie, A. O., and Maxson, R. E., Jr. (2006). Cell mixing at a neural crest-mesoderm boundary and deficient ephrin-Eph signaling in the pathogenesis of craniosynostosis. *Hum Mol Genet* **15**, pp. 1319–28.

Merzdorf, C. S. (2007). Emerging roles for zic genes in early development. *Dev Dyn* **236**, pp. 922–40.

Meyer, D., Durliat, M., Senan, F., Wolff, M., Andre, M., Hourdry, J., and Remy, P. (1997). Ets-1 and Ets-2 proto-oncogenes exhibit differential and restricted expression patterns during *Xenopus laevis* oogenesis and embryogenesis. *Int J Dev Biol* **41**, pp. 607–20.

Meyer, D., Stiegler, P., Hindelang, C., Mager, A. M., and Remy, P. (1995). Whole-mount in situ hybridization reveals the expression of the Xl-Fli gene in several lineages of migrating cells in *Xenopus* embryos. *Int J Dev Biol* **39**, pp. 909–19.

Meyer, D., Wolff, C. M., Stiegler, P., Senan, F., Befort, N., Befort, J. J., and Remy, P. (1993). Xl-fli, the *Xenopus* homologue of the fli-1 gene, is expressed during embryogenesis in a restricted pattern evocative of neural crest cell distribution. *Mech Dev* **44**, pp. 109–21.

Milewski, R. C., Chi, N. C., Li, J., Brown, C., Lu, M. M., and Epstein, J. A. (2004). Identification of minimal enhancer elements sufficient for Pax3 expression in neural crest and implication of Tead2 as a regulator of Pax3. *Development* **131**, pp. 829–37.

Miller, C. T., Swartz, M. E., Khuu, P. A., Walker, M. B., Eberhart, J. K., and Kimmel, C. B. (2007). mef2ca is required in cranial neural crest to effect Endothelin1 signaling in zebrafish. *Dev Biol* **308**, pp. 144–57.

Miller, S. J., Rangwala, F., Williams, J., Ackerman, P., Kong, S., Jegga, A. G., Kaiser, S., Aronow, B. J., Frahm, S., Kluwe, L., Mautner, V., Upadhyaya, M., Muir, D., Wallace, M., Hagen, J., Quelle, D. E., Watson, M. A., Perry, A., Gutmann, D. H., and Ratner, N. (2006). Large-scale molecular comparison of human Schwann cells to malignant peripheral nerve sheath tumor cell lines and tissues. *Cancer Res* **66**, pp. 2584–91.

Minarcik, J. C., and Golden, J. A. (2003). AP-2 and HNK-1 define distinct populations of cranial neural crest cells. *Orthod Craniofac Res* **6**, pp. 210–9.

Minchin, J. E., and Hughes, S. M. (2008). Sequential actions of Pax3 and Pax7 drive xanthophore development in zebrafish neural crest. *Dev Biol* **317**, pp. 508–22.

Mirsky, R., Woodhoo, A., Parkinson, D. B., Arthur-Farraj, P., Bhaskaran, A., and Jessen, K. R. (2008). Novel signals controlling embryonic Schwann cell development, myelination and dedifferentiation. *J Peripher Nerv Syst* **13**, pp. 122–35.

Mitchell, P. J., Timmons, P. M., Hebert, J. M., Rigby, P. W., and Tjian, R. (1991). Transcription factor AP-2 is expressed in neural crest cell lineages during mouse embryogenesis. *Genes Dev* **5**, pp. 105–19.

Mohindra, P., Zade, B., Basu, A., Patil, N., Viswanathan, S., Bakshi, A., Muckaden, M. A., and Laskar, S. (2008). Primary PNET of maxilla: an unusual presentation. *J Pediatr Hematol Oncol* **30**, pp. 474–7.

Monsoro-Burq, A. H., Wang, E., and Harland, R. (2005). Msx1 and Pax3 cooperate to mediate FGF8 and WNT signals during *Xenopus* neural crest induction. *Dev Cell* **8**, pp. 167–78.

Montelius, A., Marmigere, F., Baudet, C., Aquino, J. B., Enerback, S., and Ernfors, P. (2007). Emergence of the sensory nervous system as defined by Foxs1 expression. *Differentiation* **75**, pp. 404–17.

Montero-Balaguer, M., Lang, M. R., Sachdev, S. W., Knappmeyer, C., Stewart, R. A., De La Guardia, A., Hatzopoulos, A. K., and Knapik, E. W. (2006). The mother superior mutation ablates foxd3 activity in neural crest progenitor cells and depletes neural crest derivatives in zebrafish. *Dev Dyn* **235**, pp. 3199–212.

Morales, A. V., Perez-Alcala, S., and Barbas, J. A. (2007). Dynamic Sox5 protein expression during cranial ganglia development. *Dev Dyn* **236**, pp. 2702–7.

Morell, R., Spritz, R. A., Ho, L., Pierpont, J., Guo, W., Friedman, T. B., and Asher, J. H., Jr. (1997).

Apparent digenic inheritance of Waardenburg syndrome type 2 (WS2) and autosomal recessive ocular albinism (AROA). *Hum Mol Genet* **6**, pp. 659–64.

Moreno, T. A., and Bronner-Fraser, M. (2001). The secreted glycoprotein Noelin-1 promotes neurogenesis in *Xenopus*. *Dev Biol* **240**, pp. 340–60.

Morgan, R., and Sargent, M. G. (1997). The role in neural patterning of translation initiation factor eIF4AII; induction of neural fold genes. *Development* **124**, pp. 2751–60.

Morgan, S. C., Lee, H. Y., Relaix, F., Sandell, L. L., Levorse, J. M., and Loeken, M. R. (2008a). Cardiac outflow tract septation failure in Pax3-deficient embryos is due to p53-dependent regulation of migrating cardiac neural crest. *Mech Dev* **125**, pp. 757–67.

Morgan, S. C., Relaix, F., Sandell, L. L., and Loeken, M. R. (2008b). Oxidative stress during diabetic pregnancy disrupts cardiac neural crest migration and causes outflow tract defects. *Birth Defects Res A Clin Mol Teratol* **82**, pp. 453–63.

Mori-Akiyama, Y., Akiyama, H., Rowitch, D. H., and de Crombrugghe, B. (2003). Sox9 is required for determination of the chondrogenic cell lineage in the cranial neural crest. *Proc Natl Acad Sci U S A* **100**, pp. 9360–65.

Morikawa, Y., and Cserjesi, P. (2008). Cardiac neural crest expression of Hand2 regulates outflow and second heart field development. *Circ Res* **103**, pp. 1422–29.

Morikawa, Y., D'Autreaux, F., Gershon, M. D., and Cserjesi, P. (2007). Hand2 determines the noradrenergic phenotype in the mouse sympathetic nervous system. *Dev Biol* **307**, pp. 114–26.

Morin, S., Paradis, P., Aries, A., and Nemer, M. (2001). Serum response factor-GATA ternary complex required for nuclear signaling by a G-protein-coupled receptor. *Mol Cell Biol* **21**, pp. 1036–44.

Morriss-Kay, G. M. (1996). Craniofacial defects in AP-2 null mutant mice. *Bioessays* **18**, pp. 785–88.

Morriss-Kay, G. M., Murphy, P., Hill, R. E., and Davidson, D. R. (1991). Effects of retinoic acid excess on expression of Hox-2.9 and Krox-20 and on morphological segmentation in the hindbrain of mouse embryos. *Embo J* **10**, pp. 2985–95.

Moser, M., Ruschoff, J., and Buettner, R. (1997). Comparative analysis of AP-2 alpha and AP-2 beta gene expression during murine embryogenesis. *Dev Dyn* **208**, pp. 115–24.

Mou, Z., Tapper, A. R., and Gardner, P. D. (2009). The armadillo repeat-containing protein, ARMCX3, physically and functionally interacts with the developmental regulatory factor Sox10. *J Biol Chem* **284**, pp. 13629–40.

Mulder, G. B., Manley, N., and Maggio-Price, L. (1998). Retinoic acid-induced thymic abnormalities in the mouse are associated with altered pharyngeal morphology, thymocyte maturation defects, and altered expression of Hoxa3 and Pax1. *Teratology* **58**, pp. 263–75.

Muller, S. M., Stolt, C. C., Terszowski, G., Blum, C., Amagai, T., Kessaris, N., Iannarelli, P., Richardson, W. D., Wegner, M., and Rodewald, H. R. (2008). Neural crest origin of perivascular mesenchyme in the adult thymus. *J Immunol* **180**, pp. 5344–51.

Mummenhoff, J., Houweling, A. C., Peters, T., Christoffels, V. M., and Ruther, U. (2001). Expression of Irx6 during mouse morphogenesis. *Mech Dev* **103**, pp. 193–5.

Munchberg, S. R., and Steinbeisser, H. (1999). The *Xenopus* Ets transcription factor XER81 is a target of the FGF signaling pathway. *Mech Dev* **80**, pp. 53–65.

Murakami, H., and Arnheiter, H. (2005). Sumoylation modulates transcriptional activity of MITF in a promoter-specific manner. *Pigment Cell Res* **18**, pp. 265–77.

Murakami, M., Kataoka, K., Fukuhara, S., Nakagawa, O., and Kurihara, H. (2004a). Akt-dependent phosphorylation negatively regulates the transcriptional activity of dHAND by inhibiting the DNA binding activity. *Eur J Biochem* **271**, pp. 3330–9.

Murakami, M., Kataoka, K., Tominaga, J., Nakagawa, O., and Kurihara, H. (2004b). Differential cooperation between dHAND and three different E-proteins. *Biochem Biophys Res Commun* **323**, pp. 168–74.

Murisier, F., and Beermann, F. (2006). Genetics of pigment cells: lessons from the tyrosinase gene family. *Histol Histopathol* **21**, pp. 567–78.

Murisier, F., Guichard, S., and Beermann, F. (2006). A conserved transcriptional enhancer that specifies Tyrp1 expression to melanocytes. *Dev Biol* **298**, pp. 644–55.

Murray, S. A., and Gridley, T. (2006a). Snail1 gene function during early embryo patterning in mice. *Cell Cycle* **5**, pp. 2566–70.

Murray, S. A., and Gridley, T. (2006b). Snail family genes are required for left–right asymmetry determination, but not neural crest formation, in mice. *Proc Natl Acad Sci U S A* **103**, pp. 10300–4.

Murray, S. A., Oram, K. F., and Gridley, T. (2007). Multiple functions of Snail family genes during palate development in mice. *Development* **134**, pp. 1789–97.

Nagai, T., Aruga, J., Minowa, O., Sugimoto, T., Ohno, Y., Noda, T., and Mikoshiba, K. (2000). Zic2 regulates the kinetics of neurulation. *Proc Natl Acad Sci U S A* **97**, pp. 1618–23.

Nagase, T., Nakamura, S., Harii, K., and Osumi, N. (2001). Ectopically localized HNK-1 epitope perturbs migration of the midbrain neural crest cells in Pax6 mutant rat. *Dev Growth Differ* **43**, pp. 683–92.

Nagatomo, K., and Hashimoto, C. (2007). *Xenopus* hairy2 functions in neural crest formation by maintaining cells in a mitotic and undifferentiated state. *Dev Dyn* **236**, pp. 1475–83.

Nakada, C., Iida, A., Tabata, Y., and Watanabe, S. (2009). Forkhead transcription factor foxe1 regulates chondrogenesis in zebrafish. *J Exp Zool B Mol Dev Evol* **312**, pp. 827–40.

Nakata, K., Nagai, T., Aruga, J., and Mikoshiba, K. (1997). *Xenopus* Zic3, a primary regulator both in neural and neural crest development. *Proc Natl Acad Sci U S A* **94**, pp. 11980–5.

Nakata, K., Nagai, T., Aruga, J., and Mikoshiba, K. (1998). *Xenopus* Zic family and its role in neural and neural crest development. *Mech Dev* **75**, pp. 43–51.

Nakayama, A., Nguyen, M. T., Chen, C. C., Opdecamp, K., Hodgkinson, C. A., and Arnheiter, H. (1998). Mutations in microphthalmia, the mouse homolog of the human deafness gene MITF, affect neuroepithelial and neural crest-derived melanocytes differently. *Mech Dev* **70**, pp. 155–66.

Nakazaki, H., Reddy, A. C., Mania-Farnell, B. L., Shen, Y. W., Ichi, S., McCabe, C., George, D., McLone, D. G., Tomita, T., and Mayanil, C. S. (2008). Key basic helix–loop–helix transcription factor genes Hes1 and Ngn2 are regulated by Pax3 during mouse embryonic development. *Dev Biol* **316**, pp. 510–23.

Nakazaki, H., Shen, Y. W., Yun, B., Reddy, A., Khanna, V., Mania-Farnell, B., Ichi, S., McLone, D. G., Tomita, T., and Mayanil, C. S. (2009). Transcriptional regulation by Pax3 and TGF-beta2 signaling: a potential gene regulatory network in neural crest development. *Int J Dev Biol* **53**, pp. 69–79.

Nardelli, J., Catala, M., and Charnay, P. (2003). Establishment of embryonic neuroepithelial cell lines exhibiting an epiplastic expression pattern of region specific markers. *J Neurosci Res* **73**, pp. 737–52.

Nekrep, N., Wang, J., Miyatsuka, T., and German, M. S. (2008). Signals from the neural crest regulate beta-cell mass in the pancreas. *Development* **135**, pp. 2151–60.

Nemer, G., and Nemer, M. (2003). Transcriptional activation of BMP-4 and regulation of mammalian organogenesis by GATA-4 and -6. *Dev Biol* **254**, pp. 131–48.

Nentwich, O., Munchberg, F. E., Frommer, G., and Nordheim, A. (2001). Tissue-specific expression of the Ets gene Xsap-1 during *Xenopus laevis* development. *Mech Dev* **109**, pp. 433–6.

Newbern, J., Zhong, J., Wickramasinghe, R. S., Li, X., Wu, Y., Samuels, I., Cherosky, N., Karlo, J. C., O'Loughlin, B., Wikenheiser, J., Gargesha, M., Doughman, Y. Q., Charron, J., Ginty, D. D., Watanabe, M., Saitta, S. C., Snider, W. D., and Landreth, G. E. (2008). Mouse and human phenotypes indicate a critical conserved role for ERK2 signaling in neural crest development. *Proc Natl Acad Sci U S A* **105**, pp. 17115–20.

Ngan, E. S., Sit, F. Y., Lee, K., Miao, X., Yuan, Z., Wang, W., Nicholls, J. M., Wong, K. K., Garcia-Barcelo, M., Lui, V. C., and Tam, P. K. (2007). Implications of endocrine gland-derived vascular endothelial growth factor/prokineticin-1 signaling in human neuroblastoma progression. *Clin Cancer Res* **13**, pp. 868–75.

Nguyen, M. T., Zhu, J., Nakamura, E., Bao, X., and Mackem, S. (2009). Tamoxifen-dependent, inducible Hoxb6CreERT recombinase function in lateral plate and limb mesoderm, CNS isthmic organizer, posterior trunk neural crest, hindgut, and tailbud. *Dev Dyn* **238**, pp. 467–74.

Nguyen, V. H., Schmid, B., Trout, J., Connors, S. A., Ekker, M., and Mullins, M. C. (1998). Ventral and lateral regions of the zebrafish gastrula, including the neural crest progenitors, are established by a bmp2b/swirl pathway of genes. *Dev Biol* **199**, pp. 93–110.

Nichane, M., Ren, X., Souopgui, J., and Bellefroid, E. J. (2008). Hairy2 functions through both DNA-binding and non DNA-binding mechanisms at the neural plate border in *Xenopus*. *Dev Biol* **322**, pp. 368–80.

Nicolas, S., Massacrier, A., Caubit, X., Cau, P., and Le Parco, Y. (1996). A Distal-less-like gene is induced in the regenerating central nervous system of the urodele Pleurodeles waltl. *Mech Dev* **56**, pp. 209–20.

Niederreither, K., Vermot, J., Schuhbaur, B., Chambon, P., and Dolle, P. (2000). Retinoic acid synthesis and hindbrain patterning in the mouse embryo. *Development* **127**, pp. 75–85.

Nieto, M. A., Sargent, M. G., Wilkinson, D. G., and Cooke, J. (1994). Control of cell behavior during vertebrate development by Slug, a zinc finger gene. *Science* **264**, pp. 835–9.

Nieto, M. A., Sechrist, J., Wilkinson, D. G., and Bronner-Fraser, M. (1995). Relationship between spatially restricted Krox-20 gene expression in branchial neural crest and segmentation in the chick embryo hindbrain. *Embo J* **14**, pp. 1697–710.

Nissen, R. M., Yan, J., Amsterdam, A., Hopkins, N., and Burgess, S. M. (2003). Zebrafish foxi one modulates cellular responses to Fgf signaling required for the integrity of ear and jaw patterning. *Development* **130**, pp. 2543–54.

Nonaka, D., Chiriboga, L., and Rubin, B. P. (2008). Sox10: a pan-schwannian and melanocytic marker. *Am J Surg Pathol* **32**, pp. 1291–8.

Nonchev, S., Vesque, C., Maconochie, M., Seitanidou, T., Ariza-McNaughton, L., Frain, M., Marshall, H., Sham, M. H., Krumlauf, R., and Charnay, P. (1996). Segmental expression of Hoxa-2 in the hindbrain is directly regulated by Krox-20. *Development* **122**, pp. 543–54.

O'Brien, E. K., d'Alencon, C., Bonde, G., Li, W., Schoenebeck, J., Allende, M. L., Gelb, B. D., Yelon, D., Eisen, J. S., and Cornell, R. A. (2004). Transcription factor Ap-2alpha is necessary for development of embryonic melanophores, autonomic neurons and pharyngeal skeleton in zebrafish. *Dev Biol* **265**, pp. 246–61.

O'Donnell, M., Hong, C. S., Huang, X., Delnicki, R. J., and Saint-Jeannet, J. P. (2006). Functional analysis of Sox8 during neural crest development in *Xenopus*. *Development* **133**, pp. 3817–26.

Oh, J., Richardson, J. A., and Olson, E. N. (2005). Requirement of myocardin-related transcription factor-B for remodeling of branchial arch arteries and smooth muscle differentiation. *Proc Natl Acad Sci U S A* **102**, pp. 15122–7.

Ohira, M., Morohashi, A., Inuzuka, H., Shishikura, T., Kawamoto, T., Kageyama, H., Nakamura, Y., Isogai, E., Takayasu, H., Sakiyama, S., Suzuki, Y., Sugano, S., Goto, T., Sato, S., and

Nakagawara, A. (2003). Expression profiling and characterization of 4200 genes cloned from primary neuroblastomas: identification of 305 genes differentially expressed between favorable and unfavorable subsets. *Oncogene* **22**, pp. 5525–36.

Okajima, K., Paznekas, W. A., Burstyn, T., and Jabs, E. W. (2001). Polymorphisms in the Human SNAIL (SNAI1) gene. *Mol Cell Probes* **15**, pp. 53–5.

Okamura, Y., and Saga, Y. (2008). Notch signaling is required for the maintenance of enteric neural crest progenitors. *Development* **135**, pp. 3555–65.

Opdecamp, K., Nakayama, A., Nguyen, M. T., Hodgkinson, C. A., Pavan, W. J., and Arnheiter, H. (1997). Melanocyte development in vivo and in neural crest cell cultures: crucial dependence on the Mitf basic-helix-loop-helix-zipper transcription factor. *Development* **124**, pp. 2377–86.

Oram, K. F., Carver, E. A., and Gridley, T. (2003). Slug expression during organogenesis in mice. *Anat Rec A Discov Mol Cell Evol Biol* **271**, pp. 189–91.

Ormestad, M., Astorga, J., and Carlsson, P. (2004). Differences in the embryonic expression patterns of mouse Foxf1 and -2 match their distinct mutant phenotypes. *Dev Dyn* **229**, pp. 328–33.

O'Rourke, M. P., and Tam, P. P. (2002). Twist functions in mouse development. *Int J Dev Biol* **46**, pp. 401–13.

Osorio, L., Teillet, M. A., Palmeirim, I., and Catala, M. (2009). Neural crest ontogeny during secondary neurulation: a gene expression pattern study in the chick embryo. *Int J Dev Biol* **53**, pp. 641–8.

Osumi-Yamashita, N., Kuratani, S., Ninomiya, Y., Aoki, K., Iseki, S., Chareonvit, S., Doi, H., Fujiwara, M., Watanabe, T., and Eto, K. (1997). Cranial anomaly of homozygous rSey rat is associated with a defect in the migration pathway of midbrain crest cells. *Dev Growth Differ* **39**, pp. 53–67.

Ota, M., and Ito, K. (2003). Induction of neurogenin-1 expression by sonic hedgehog: Its role in development of trigeminal sensory neurons. *Dev Dyn* **227**, pp. 544–51.

Ota, M., and Ito, K. (2006). BMP and FGF-2 regulate neurogenin-2 expression and the differentiation of sensory neurons and glia. *Dev Dyn* **235**, pp. 646–55.

Ota, M. S., Loebel, D. A., O'Rourke, M. P., Wong, N., Tsoi, B., and Tam, P. P. (2004). Twist is required for patterning the cranial nerves and maintaining the viability of mesodermal cells. *Dev Dyn* **230**, pp. 216–28.

Oxtoby, E., and Jowett, T. (1993). Cloning of the zebrafish krox-20 gene (krx-20) and its expression during hindbrain development. *Nucleic Acids Res* **21**, pp. 1087–95.

Palmer, M. B., Majumder, P., Cooper, J. C., Yoon, H., Wade, P. A., and Boss, J. M. (2009). Yin yang 1 regulates the expression of snail through a distal enhancer. *Mol Cancer Res* **7**, pp. 221–9.

Pani, L., Horal, M., and Loeken, M. R. (2002). Rescue of neural tube defects in Pax-3-deficient embryos by p53 loss of function: implications for Pax-3- dependent development and tumorigenesis. *Genes Dev* **16**, pp. 676–80.

Panteleyev, A. A., Mitchell, P. J., Paus, R., and Christiano, A. M. (2003). Expression patterns of the transcription factor AP-2alpha during hair follicle morphogenesis and cycling. *J Invest Dermatol* **121**, pp. 13–9.

Papis, E., Bernardini, G., Gornati, R., Menegola, E., and Prati, M. (2007). Gene expression in *Xenopus laevis* embryos after Triadimefon exposure. *Gene Expr Patterns* **7**, pp. 137–42.

Paratore, C., Brugnoli, G., Lee, H. Y., Suter, U., and Sommer, L. (2002a). The role of the Ets domain transcription factor Erm in modulating differentiation of neural crest stem cells. *Dev Biol* **250**, pp. 168–80.

Paratore, C., Eichenberger, C., Suter, U., and Sommer, L. (2002b). Sox10 haploinsufficiency affects maintenance of progenitor cells in a mouse model of Hirschsprung disease. *Hum Mol Genet* **11**, pp. 3075–85.

Paratore, C., Goerich, D. E., Suter, U., Wegner, M., and Sommer, L. (2001). Survival and glial fate acquisition of neural crest cells are regulated by an interplay between the transcription factor Sox10 and extrinsic combinatorial signaling. *Development* **128**, pp. 3949–61.

Parisi, M. A., and Kapur, R. P. (2000). Genetics of Hirschsprung disease. *Curr Opin Pediatr* **12**, pp. 610–7.

Park, I. H., Zhao, R., West, J. A., Yabuuchi, A., Huo, H., Ince, T. A., Lerou, P. H., Lensch, M. W., and Daley, G. Q. (2008). Reprogramming of human somatic cells to pluripotency with defined factors. *Nature* **451**, pp. 141–6.

Parker, C. J., Shawcross, S. G., Li, H., Wang, Q. Y., Herrington, C. S., Kumar, S., MacKie, R. M., Prime, W., Rennie, I. G., Sisley, K., and Kumar, P. (2004). Expression of PAX 3 alternatively spliced transcripts and identification of two new isoforms in human tumors of neural crest origin. *Int J Cancer* **108**, pp. 314–20.

Passeron, T., Valencia, J. C., Bertolotto, C., Hoashi, T., Le Pape, E., Takahashi, K., Ballotti, R., and Hearing, V. J. (2007). SOX9 is a key player in ultraviolet B-induced melanocyte differentiation and pigmentation. *Proc Natl Acad Sci U S A* **104**, pp. 13984–9.

Pattyn, A., Morin, X., Cremer, H., Goridis, C., and Brunet, J. F. (1999). The homeobox gene Phox2b is essential for the development of autonomic neural crest derivatives. *Nature* **399**, pp. 366–70.

Payson, R. A., Chotani, M. A., and Chiu, I. M. (1998). Regulation of a promoter of the fibroblast growth factor 1 gene in prostate and breast cancer cells. *J Steroid Biochem Mol Biol* **66**, pp. 93–103.

Paznekas, W. A., Okajima, K., Schertzer, M., Wood, S., and Jabs, E. W. (1999). Genomic organization, expression, and chromosome location of the human SNAIL gene (SNAI1) and a related processed pseudogene (SNAI1P). *Genomics* **62**, pp. 42–9.

Peirano, R. I., and Wegner, M. (2000). The glial transcription factor Sox10 binds to DNA both as monomer and dimer with different functional consequences. *Nucleic Acids Res* **28**, pp. 3047–55.

Pennati, R., Groppelli, S., de Bernardi, F., and Sotgia, C. (2001). Action of valproic acid on *Xenopus laevis* development: teratogenic effects on eyes. *Teratog Carcinog Mutagen* **21**, pp. 121–33.

Pera, E., and Kessel, M. (1999). Expression of DLX3 in chick embryos. *Mech Dev* **89**, pp. 189–93.

Perez, S. E., Rebelo, S., and Anderson, D. J. (1999). Early specification of sensory neuron fate revealed by expression and function of neurogenins in the chick embryo. *Development* **126**, pp. 1715–28.

Perez-Alcala, S., Nieto, M. A., and Barbas, J. A. (2004). LSox5 regulates RhoB expression in the neural tube and promotes generation of the neural crest. *Development* **131**, pp. 4455–65.

Perez-Mancera, P. A., Gonzalez-Herrero, I., Maclean, K., Turner, A. M., Yip, M. Y., Sanchez-Martin, M., Garcia, J. L., Robledo, C., Flores, T., Gutierrez-Adan, A., Pintado, B., and Sanchez-Garcia, I. (2006). SLUG (SNAI2) overexpression in embryonic development. *Cytogenet Genome Res* **114**, pp. 24–9.

Perez-Mancera, P. A., Gonzalez-Herrero, I., Perez-Caro, M., Gutierrez-Cianca, N., Flores, T., Gutierrez-Adan, A., Pintado, B., Sanchez-Martin, M., and Sanchez-Garcia, I. (2005). SLUG in cancer development. *Oncogene* **24**, pp. 3073–82.

Perez-Moreno, M. A., Locascio, A., Rodrigo, I., Dhondt, G., Portillo, F., Nieto, M. A., and Cano, A. (2001). A new role for E12/E47 in the repression of E-cadherin expression and epithelial–mesenchymal transitions. *J Biol Chem* **276**, pp. 27424–31.

Peters, H., Neubuser, A., and Balling, R. (1998). Pax genes and organogenesis: Pax9 meets tooth development. *Eur J Oral Sci* **106 Suppl 1**, pp. 38–43.

Pfisterer, P., Ehlermann, J., Hegen, M., and Schorle, H. (2002). A subtractive gene expression screen suggests a role of transcription factor AP-2 alpha in control of proliferation and differentiation. *J Biol Chem* **277**, pp. 6637–44.

Pham, V. N., Lawson, N. D., Mugford, J. W., Dye, L., Castranova, D., Lo, B., and Weinstein, B. M. (2007). Combinatorial function of ETS transcription factors in the developing vasculature. *Dev Biol* **303**, pp. 772–83.

Pierpont, M. E., Basson, C. T., Benson, D. W., Jr., Gelb, B. D., Giglia, T. M., Goldmuntz, E., McGee, G., Sable, C. A., Srivastava, D., and Webb, C. L. (2007). Genetic basis for congenital heart defects: current knowledge: a scientific statement from the American Heart Association Congenital Cardiac Defects Committee, Council on Cardiovascular Disease in the Young: endorsed by the American Academy of Pediatrics. *Circulation* **115**, pp. 3015–38.

Pietras, A., Gisselsson, D., Ora, I., Noguera, R., Beckman, S., Navarro, S., and Pahlman, S. (2008). High levels of HIF-2alpha highlight an immature neural crest-like neuroblastoma cell cohort located in a perivascular niche. *J Pathol* **214**, pp. 482–8.

Pietsch, J., Delalande, J. M., Jakaitis, B., Stensby, J. D., Dohle, S., Talbot, W. S., Raible, D. W., and Shepherd, I. T. (2006). lessen encodes a zebrafish trap100 required for enteric nervous system development. *Development* **133**, pp. 395–406.

Pilon, N., Raiwet, D., Viger, R. S., and Silversides, D. W. (2008). Novel pre- and post-gastrulation expression of Gata4 within cells of the inner cell mass and migratory neural crest cells. *Dev Dyn* **237**, pp. 1133–43.

Pingault, V., Bondurand, N., Kuhlbrodt, K., Goerich, D. E., Prehu, M. O., Puliti, A., Herbarth, B., Hermans-Borgmeyer, I., Legius, E., Matthijs, G., Amiel, J., Lyonnet, S., Ceccherini, I., Romeo, G., Smith, J. C., Read, A. P., Wegner, M., and Goossens, M. (1998). SOX10 mutations in patients with Waardenburg–Hirschsprung disease. *Nat Genet* **18**, pp. 171–3.

Pingault, V., Guiochon-Mantel, A., Bondurand, N., Faure, C., Lacroix, C., Lyonnet, S., Goossens, M., and Landrieu, P. (2000). Peripheral neuropathy with hypomyelination, chronic intestinal pseudo-obstruction and deafness: a developmental "neural crest syndrome" related to a SOX10 mutation. *Ann Neurol* **48**, pp. 671–6.

Pisano, J. M., Colon-Hastings, F., and Birren, S. J. (2000). Postmigratory enteric and sympathetic neural precursors share common, developmentally regulated, responses to BMP2. *Dev Biol* **227**, pp. 1–11.

Pitera, J. E., Smith, V. V., Thorogood, P., and Milla, P. J. (1999). Coordinated expression of 3' hox genes during murine embryonal gut development: an enteric Hox code. *Gastroenterology* **117**, pp. 1339–51.

Planque, N., Raposo, G., Leconte, L., Anezo, O., Martin, P., and Saule, S. (2004). Microphthalmia transcription factor induces both retinal pigmented epithelium and neural crest melanocytes from neuroretina cells. *J Biol Chem* **279**, pp. 41911–7.

Plaza Menacho, I., Koster, R., van der Sloot, A. M., Quax, W. J., Osinga, J., van der Sluis, T., Hollema, H., Burzynski, G. M., Gimm, O., Buys, C. H., Eggen, B. J., and Hofstra, R. M. (2005). RET-familial medullary thyroid carcinoma mutants Y791F and S891A activate a Src/JAK/STAT3 pathway, independent of glial cell line-derived neurotrophic factor. *Cancer Res* **65**, pp. 1729–37.

Pogoda, H. M., von der Hardt, S., Herzog, W., Kramer, C., Schwarz, H., and Hammerschmidt, M. (2006). The proneural gene ascl1a is required for endocrine differentiation and cell survival in the zebrafish adenohypophysis. *Development* **133**, pp. 1079–89.

Pohl, B. S., and Knochel, W. (2001). Overexpression of the transcriptional repressor FoxD3 prevents neural crest formation in *Xenopus* embryos. *Mech Dev* **103**, pp. 93–106.

Pohl, B. S., and Knochel, W. (2002). Temporal and spatial expression patterns of FoxD2 during the early development of *Xenopus laevis*. *Mech Dev* **111**, pp. 181–4.

Pohl, B. S., Schon, C., Rossner, A., and Knochel, W. (2004). The FoxO-subclass in *Xenopus laevis* development. *Gene Expr Patterns* **5**, pp. 187–92.

Pomp, O., Brokhman, I., Ben-Dor, I., Reubinoff, B., and Goldstein, R. S. (2005). Generation of peripheral sensory and sympathetic neurons and neural crest cells from human embryonic stem cells. *Stem Cells* **23**, pp. 923–30.

Porras, D., and Brown, C. B. (2008). Temporal-spatial ablation of neural crest in the mouse results in cardiovascular defects. *Dev Dyn* **237**, pp. 153–62.

Potterf, S. B., Furumura, M., Dunn, K. J., Arnheiter, H., and Pavan, W. J. (2000). Transcription factor hierarchy in Waardenburg syndrome: regulation of MITF expression by SOX10 and PAX3. *Hum Genet* **107**, pp. 1–6.

Potterf, S. B., Mollaaghababa, R., Hou, L., Southard-Smith, E. M., Hornyak, T. J., Arnheiter, H., and Pavan, W. J. (2001). Analysis of SOX10 function in neural crest-derived melanocyte development: SOX10-dependent transcriptional control of dopachrome tautomerase. *Dev Biol* **237**, pp. 245–57.

Prince, S., Wiggins, T., Hulley, P. A., and Kidson, S. H. (2003). Stimulation of melanogenesis by tetradecanoylphorbol 13-acetate (TPA) in mouse melanocytes and neural crest cells. *Pigment Cell Res* **16**, pp. 26–34.

Prince, V. E., Moens, C. B., Kimmel, C. B., and Ho, R. K. (1998). Zebrafish hox genes: expression in the hindbrain region of wild-type and mutants of the segmentation gene, valentino. *Development* **125**, pp. 393–406.

Pruitt, S. C., Bussman, A., Maslov, A. Y., Natoli, T. A., and Heinaman, R. (2004). Hox/Pbx and Brn binding sites mediate Pax3 expression in vitro and in vivo. *Gene Expr Patterns* **4**, pp. 671–85.

Ptok, M., and Morlot, S. (2006). [Unilateral sensineural deafness associated with mutations in the PAX3-gene in Waardenburg syndrome type I]. *Hno* **54**, pp. 557–60.

Puri, P., and Shinkai, T. (2004). Pathogenesis of Hirschsprung's disease and its variants: recent progress. *Semin Pediatr Surg* **13**, pp. 18–24.

Qiu, Y., Pereira, F. A., DeMayo, F. J., Lydon, J. P., Tsai, S. Y., and Tsai, M. J. (1997). Null mutation of mCOUP-TFI results in defects in morphogenesis of the glossopharyngeal ganglion, axonal projection, and arborization. *Genes Dev* **11**, pp. 1925–37.

Qu, S., Tucker, S. C., Zhao, Q., deCrombrugghe, B., and Wisdom, R. (1999). Physical and genetic interactions between Alx4 and Cart1. *Development* **126**, pp. 359–69.

Raible, F., and Brand, M. (2001). Tight transcriptional control of the ETS domain factors Erm and Pea3 by Fgf signaling during early zebrafish development. *Mech Dev* **107**, pp. 105–17.

Raid, R., Krinka, D., Bakhoff, L., Abdelwahid, E., Jokinen, E., Karner, M., Malva, M., Meier, R., Pelliniemi, L. J., Ploom, M., Sizarov, A., Pooga, M., and Karis, A. (2009). Lack of Gata3 results in conotruncal heart anomalies in mouse. *Mech Dev* **126**, pp. 80–9.

Rajan, P., Panchision, D. M., Newell, L. F., and McKay, R. D. (2003). BMPs signal alternately through a SMAD or FRAP-STAT pathway to regulate fate choice in CNS stem cells. *J Cell Biol* **161**, pp. 911–21.

Rau, M. J., Fischer, S., and Neumann, C. J. (2006). Zebrafish Trap230/Med12 is required as a coactivator for Sox9-dependent neural crest, cartilage and ear development. *Dev Biol* **296**, pp. 83–93.

Rausa, F. M., Galarneau, L., Belanger, L., and Costa, R. H. (1999). The nuclear receptor fetoprotein transcription factor is coexpressed with its target gene HNF-3beta in the developing murine liver, intestine and pancreas. *Mech Dev* **89**, pp. 185–8.

Read, A. P., and Newton, V. E. (1997). Waardenburg syndrome. *J Med Genet* **34**, pp. 656–65.

Reamon-Buettner, S. M., Ciribilli, Y., Inga, A., and Borlak, J. (2008). A loss-of-function mutation in the binding domain of HAND1 predicts hypoplasia of the human hearts. *Hum Mol Genet* **17**, pp. 1397–405.

Rehberg, S., Lischka, P., Glaser, G., Stamminger, T., Wegner, M., and Rosorius, O. (2002). Sox10 is an active nucleocytoplasmic shuttle protein, and shuttling is crucial for Sox10-mediated transactivation. *Mol Cell Biol* **22**, pp. 5826–34.

Reichenbach, B., Delalande, J. M., Kolmogorova, E., Prier, A., Nguyen, T., Smith, C. M., Holzschuh, J., and Shepherd, I. T. (2008). Endoderm-derived Sonic hedgehog and mesoderm Hand2 expression are required for enteric nervous system development in zebrafish. *Dev Biol* **318**, pp. 52–64.

Reiprich, S., Stolt, C. C., Schreiner, S., Parlato, R., and Wegner, M. (2008). SoxE proteins are differentially required in mouse adrenal gland development. *Mol Biol Cell* **19**, pp. 1575–86.

Remy, P., and Baltzinger, M. (2000). The Ets-transcription factor family in embryonic development: lessons from the amphibian and bird. *Oncogene* **19**, pp. 6417–31.

Restivo, A., Piacentini, G., Placidi, S., Saffirio, C., and Marino, B. (2006). Cardiac outflow tract: a review of some embryogenetic aspects of the conotruncal region of the heart. *Anat Rec A Discov Mol Cell Evol Biol* **288**, pp. 936–43.

Rhim, H., Savagner, P., Thibaudeau, G., Thiery, J. P., and Pavan, W. J. (1997). Localization of a neural crest transcription factor, Slug, to mouse chromosome 16 and human chromosome 8. *Mamm Genome* **8**, pp. 872–3.

Riley, P., Anson-Cartwright, L., and Cross, J. C. (1998). The Hand1 bHLH transcription factor is essential for placentation and cardiac morphogenesis. *Nat Genet* **18**, pp. 271–5.

Riley, P. R., Gertsenstein, M., Dawson, K., and Cross, J. C. (2000). Early exclusion of hand1-deficient cells from distinct regions of the left ventricular myocardium in chimeric mouse embryos. *Dev Biol* **227**, pp. 156–68.

Rodger, J., Ziman, M. R., Papadimitriou, J. M., and Kay, P. H. (1999). Pax7 is expressed in the capsules surrounding adult mouse neuromuscular spindles. *Biochem Cell Biol* **77**, pp. 153–6.

Rodrigues, C. O., Nerlick, S. T., White, E. L., Cleveland, J. L., and King, M. L. (2008). A Myc-Slug (Snail2)/Twist regulatory circuit directs vascular development. *Development* **135**, pp. 1903–11.

Rogers, C. D., Harafuji, N., Archer, T., Cunningham, D. D., and Casey, E. S. (2009). *Xenopus* Sox3 activates sox2 and geminin and indirectly represses Xvent2 expression to induce neural progenitor formation at the expense of non-neural ectodermal derivatives. *Mech Dev* **126**, pp. 42–55.

Rorie, C. J., Thomas, V. D., Chen, P., Pierce, H. H., O'Bryan, J. P., and Weissman, B. E. (2004). The Ews/Fli-1 fusion gene switches the differentiation program of neuroblastomas to Ewing sarcoma/peripheral primitive neuroectodermal tumors. *Cancer Res* **64**, pp. 1266–77.

Rothhammer, T., Hahne, J. C., Florin, A., Poser, I., Soncin, F., Wernert, N., and Bosserhoff, A. K. (2004). The Ets-1 transcription factor is involved in the development and invasion of malignant melanoma. *Cell Mol Life Sci* **61**, pp. 118–28.

Rowe, A., and Brickell, P. M. (1995). Expression of the chicken retinoid X receptor-gamma gene in migrating cranial neural crest cells. *Anat Embryol (Berl)* **192**, pp. 1–8.

Rowe, A., Eager, N. S., and Brickell, P. M. (1991). A member of the RXR nuclear receptor family is expressed in neural-crest-derived cells of the developing chick peripheral nervous system. *Development* **111**, pp. 771–8.

Roy, S., and Ng, T. (2004). Blimp-1 specifies neural crest and sensory neuron progenitors in the zebrafish embryo. *Curr Biol* **14**, pp. 1772–7.

Ruberte, E., Dolle, P., Chambon, P., and Morriss-Kay, G. (1991). Retinoic acid receptors and cellular retinoid binding proteins. II. Their differential pattern of transcription during early morphogenesis in mouse embryos. *Development* **111**, pp. 45–60.

Ruberte, E., Wood, H. B., and Morriss-Kay, G. M. (1997). Prorhombomeric subdivision of the mammalian embryonic hindbrain: is it functionally meaningful? *Int J Dev Biol* **41**, pp. 213–22.

Ruest, L. B., Dager, M., Yanagisawa, H., Charite, J., Hammer, R. E., Olson, E. N., Yanagisawa, M., and Clouthier, D. E. (2003). dHAND-Cre transgenic mice reveal specific potential functions of dHAND during craniofacial development. *Dev Biol* **257**, pp. 263–77.

Ruest, L. B., Xiang, X., Lim, K. C., Levi, G., and Clouthier, D. E. (2004). Endothelin-A receptor-dependent and -independent signaling pathways in establishing mandibular identity. *Development* **131**, pp. 4413–23.

Sahar, D. E., Longaker, M. T., and Quarto, N. (2005). Sox9 neural crest determinant gene controls patterning and closure of the posterior frontal cranial suture. *Dev Biol* **280**, pp. 344–61.

Saint-Jeannet, J. P., He, X., Varmus, H. E., and Dawid, I. B. (1997). Regulation of dorsal fate in the neuraxis by Wnt-1 and Wnt-3a. *Proc Natl Acad Sci U S A* **94**, pp. 13713–8.

Saito, Y., Gotoh, M., Ujiie, Y., Izutsu, Y., and Maeno, M. (2009). Involvement of AP-2rep in morphogenesis of the axial mesoderm in *Xenopus* embryo. *Cell Tissue Res* **335**, pp. 357–69.

Sakai, D., Suzuki, T., Osumi, N., and Wakamatsu, Y. (2006). Cooperative action of Sox9, Snail2 and PKA signaling in early neural crest development. *Development* **133**, pp. 1323–33.

Sakai, D., Tanaka, Y., Endo, Y., Osumi, N., Okamoto, H., and Wakamatsu, Y. (2005). Regulation of Slug transcription in embryonic ectoderm by beta-catenin-Lef/Tcf and BMP-Smad signaling. *Dev Growth Differ* **47**, pp. 471–82.

Saldivar, J. R., Sechrist, J. W., Krull, C. E., Ruffins, S., and Bronner-Fraser, M. (1997). Dorsal hindbrain ablation results in rerouting of neural crest migration and changes in gene expression, but normal hyoid development. *Development* **124**, pp. 2729–39.

Sanchez-Martin, M., Perez-Losada, J., Rodriguez-Garcia, A., Gonzalez-Sanchez, B., Korf, B. R., Kuster, W., Moss, C., Spritz, R. A., and Sanchez-Garcia, I. (2003). Deletion of the SLUG (SNAI2) gene results in human piebaldism. *Am J Med Genet A* **122A**, pp. 125–32.

Sanchez-Martin, M., Rodriguez-Garcia, A., Perez-Losada, J., Sagrera, A., Read, A. P., and Sanchez-Garcia, I. (2002). SLUG (SNAI2) deletions in patients with Waardenburg disease. *Hum Mol Genet* **11**, pp. 3231–6.

Sasai, N., Mizuseki, K., and Sasai, Y. (2001). Requirement of FoxD3-class signaling for neural crest determination in *Xenopus*. *Development* **128**, pp. 2525–36.

Sato, T., Kurihara, Y., Asai, R., Kawamura, Y., Tonami, K., Uchijima, Y., Heude, E., Ekker, M., Levi, G., and Kurihara, H. (2008). An endothelin-1 switch specifies maxillomandibular identity. *Proc Natl Acad Sci U S A* **105**, pp. 18806–11.

Sato, T., Sasai, N., and Sasai, Y. (2005). Neural crest determination by co-activation of Pax3 and Zic1 genes in *Xenopus* ectoderm. *Development* **132**, pp. 2355–63.

Sauka-Spengler, T., and Bronner-Fraser, M. (2008). A gene regulatory network orchestrates neural crest formation. *Nat Rev Mol Cell Biol* **9**, pp. 557–68.

Savagner, P., Karavanova, I., Perantoni, A., Thiery, J. P., and Yamada, K. M. (1998). Slug mRNA is expressed by specific mesodermal derivatives during rodent organogenesis. *Dev Dyn* **213**, pp. 182–7.

Scemama, J. L., Hunter, M., McCallum, J., Prince, V., and Stellwag, E. (2002). Evolutionary divergence of vertebrate Hoxb2 expression patterns and transcriptional regulatory loci. *J Exp Zool* **294**, pp. 285–99.

Schafer, K., Neuhaus, P., Kruse, J., and Braun, T. (2003). The homeobox gene Lbx1 specifies a subpopulation of cardiac neural crest necessary for normal heart development. *Circ Res* **92**, pp. pp. 73–80.

Schepers, G. E., Bullejos, M., Hosking, B. M., and Koopman, P. (2000). Cloning and characterisation of the Sry-related transcription factor gene Sox8. *Nucleic Acids Res* **28**, pp. 1473–80.

Schlierf, B., Lang, S., Kosian, T., Werner, T., and Wegner, M. (2005). The high-mobility group transcription factor Sox10 interacts with the N-myc-interacting protein Nmi. *J Mol Biol* **353**, pp. 1033–42.

Schlierf, B., Ludwig, A., Klenovsek, K., and Wegner, M. (2002). Cooperative binding of Sox10 to DNA: requirements and consequences. *Nucleic Acids Res* **30**, pp. 5509–16.

Schneider, C., Wicht, H., Enderich, J., Wegner, M., and Rohrer, H. (1999). Bone morphogenetic proteins are required in vivo for the generation of sympathetic neurons. *Neuron* **24**, pp. 861–70.

Scholl, F. A., Kamarashev, J., Murmann, O. V., Geertsen, R., Dummer, R., and Schafer, B. W. (2001). PAX3 is expressed in human melanomas and contributes to tumor cell survival. *Cancer Res* **61**, pp. 823–6.

Schorle, H., Meier, P., Buchert, M., Jaenisch, R., and Mitchell, P. J. (1996). Transcription factor AP-2 essential for cranial closure and craniofacial development. *Nature* **381**, pp. 235–8.

Schreiner, S., Cossais, F., Fischer, K., Scholz, S., Bosl, M. R., Holtmann, B., Sendtner, M., and Wegner, M. (2007). Hypomorphic Sox10 alleles reveal novel protein functions and unravel developmental differences in glial lineages. *Development* **134**, pp. 3271–81.

Schuff, M., Rossner, A., Donow, C., and Knochel, W. (2006). Temporal and spatial expression patterns of FoxN genes in *Xenopus laevis* embryos. *Int J Dev Biol* **50**, pp. 429–34.

Schuff, M., Rossner, A., Wacker, S. A., Donow, C., Gessert, S., and Knochel, W. (2007). FoxN3 is required for craniofacial and eye development of *Xenopus laevis*. *Dev Dyn* **236**, pp. 226–39.

Schulte, J. H., Kirfel, J., Lim, S., Schramm, A., Friedrichs, N., Deubzer, H. E., Witt, O., Eggert, A., and Buettner, R. (2008). Transcription factor AP2alpha (TFAP2a) regulates differentiation and proliferation of neuroblastoma cells. *Cancer Lett* **271**, pp. 56–63.

Schulte, T. W., Toretsky, J. A., Ress, E., Helman, L., and Neckers, L. M. (1997). Expression of PAX3 in Ewing's sarcoma family of tumors. *Biochem Mol Med* **60**, pp. 121–6.

Schulte-Merker, S., Hammerschmidt, M., Beuchle, D., Cho, K. W., De Robertis, E. M., and Nusslein-Volhard, C. (1994). Expression of zebrafish goosecoid and no tail gene products in wild-type and mutant no tail embryos. *Development* **120**, pp. 843–52.

Sechrist, J., Nieto, M. A., Zamanian, R. T., and Bronner-Fraser, M. (1995). Regulative response of the cranial neural tube after neural fold ablation: spatiotemporal nature of neural crest regeneration and up-regulation of Slug. *Development* **121**, pp. 4103–15.

Sefton, M., Sanchez, S., and Nieto, M. A. (1998). Conserved and divergent roles for members of the Snail family of transcription factors in the chick and mouse embryo. *Development* **125**, pp. 3111–21.

Sekiya, T., Muthurajan, U. M., Luger, K., Tulin, A. V., and Zaret, K. S. (2009). Nucleosome-binding affinity as a primary determinant of the nuclear mobility of the pioneer transcription factor FoxA. *Genes Dev* **23**, pp. 804–9.

Sela-Donenfeld, D., and Kalcheim, C. (1999). Regulation of the onset of neural crest migration by coordinated activity of BMP4 and Noggin in the dorsal neural tube. *Development* **126**, pp. 4749–62.

Selleck, M. A., Garcia-Castro, M. I., Artinger, K. B., and Bronner-Fraser, M. (1998). Effects of Shh and Noggin on neural crest formation demonstrate that BMP is required in the neural tube but not ectoderm. *Development* **125**, pp. 4919–30.

Semba, I., Nonaka, K., Takahashi, I., Takahashi, K., Dashner, R., Shum, L., Nuckolls, G. H., and Slavkin, H. C. (2000). Positionally-dependent chondrogenesis induced by BMP4 is co-regulated by Sox9 and Msx2. *Dev Dyn* **217**, pp. 401–14.

Seo, H. C., Saetre, B. O., Havik, B., Ellingsen, S., and Fjose, A. (1998). The zebrafish Pax3 and Pax7 homologues are highly conserved, encode multiple isoforms and show dynamic segment-like expression in the developing brain. *Mech Dev* **70**, pp. 49–63.

Seo, S., and Kume, T. (2006). Forkhead transcription factors, Foxc1 and Foxc2, are required for the morphogenesis of the cardiac outflow tract. *Dev Biol* **296**, pp. 421–36.

Serbedzija, G. N., and McMahon, A. P. (1997). Analysis of neural crest cell migration in Splotch mice using a neural crest-specific LacZ reporter. *Dev Biol* **185**, pp. 139–47.

Sers, C., Kirsch, K., Rothbacher, U., Riethmuller, G., and Johnson, J. P. (1993). Genomic organization of the melanoma-associated glycoprotein MUC18: implications for the evolution of the immunoglobulin domains. *Proc Natl Acad Sci U S A* **90**, pp. 8514–8.

Seufert, D. W., Hegde, R. S., Nekkalapudi, S., Kelly, L. E., and El-Hodiri, H. M. (2005). Expression of a novel Ski-like gene in *Xenopus* development. *Gene Expr Patterns* **6**, pp. 22–8.

Sham, M. H., Hunt, P., Nonchev, S., Papalopulu, N., Graham, A., Boncinelli, E., and Krumlauf, R. (1992). Analysis of the murine Hox-2.7 gene: conserved alternative transcripts with differential distributions in the nervous system and the potential for shared regulatory regions. *Embo J* **11**, pp. 1825–36.

Sharpe, C., and Goldstone, K. (2000). Retinoid signalling acts during the gastrula stages to promote primary neurogenesis. *Int J Dev Biol* **44**, pp. 463–70.

Shen, H., Wilke, T., Ashique, A. M., Narvey, M., Zerucha, T., Savino, E., Williams, T., and Richman, J. M. (1997). Chicken transcription factor AP-2: cloning, expression and its role in outgrowth of facial prominences and limb buds. *Dev Biol* **188**, pp. 248–66.

Shoba, T., Dheen, S. T., and Tay, S. S. (2002). Retinoic acid influences the expression of the neuronal regulatory genes Mash-1 and c-ret in the developing rat heart. *Neurosci Lett* **318**, pp. 129–32.

Shows, K. H., and Shiang, R. (2008). Regulation of the mouse Treacher Collins syndrome homolog (Tcof1) promoter through differential repression of constitutive expression. *DNA Cell Biol* **27**, pp. 589–600.

Sieber-Blum, M., and Hu, Y. (2008). Mouse epidermal neural crest stem cell (EPI-NCSC) cultures. *J Vis Exp*.

Sieber-Blum, M., Ito, K., Richardson, M. K., Langtimm, C. J., and Duff, R. S. (1993). Distribution of pluripotent neural crest cells in the embryo and the role of brain-derived neurotrophic factor in the commitment to the primary sensory neuron lineage. *J Neurobiol* **24**, pp. 173–84.

Sieber-Blum, M., Schnell, L., Grim, M., Hu, Y. F., Schneider, R., and Schwab, M. E. (2006). Characterization of epidermal neural crest stem cell (EPI-NCSC) grafts in the lesioned spinal cord. *Mol Cell Neurosci* **32**, pp. 67–81.

Smith, R. S., Zabaleta, A., Kume, T., Savinova, O. V., Kidson, S. H., Martin, J. E., Nishimura, D. Y., Alward, W. L., Hogan, B. L., and John, S. W. (2000). Haploinsufficiency of the transcription factors FOXC1 and FOXC2 results in aberrant ocular development. *Hum Mol Genet* **9**, pp. 1021–32.

Smits, P., Li, P., Mandel, J., Zhang, Z., Deng, J. M., Behringer, R. R., de Crombrugghe, B., and Lefebvre, V. (2001). The transcription factors L-Sox5 and Sox6 are essential for cartilage formation. *Dev Cell* **1**, pp. 277–90.

Sock, E., Schmidt, K., Hermanns-Borgmeyer, I., Bosl, M. R., and Wegner, M. (2001). Idiopathic weight reduction in mice deficient in the high-mobility-group transcription factor Sox8. *Mol Cell Biol* **21**, pp. 6951–59.

Solloway, M. J., and Robertson, E. J. (1999). Early embryonic lethality in Bmp5;Bmp7 double mutant mice suggests functional redundancy within the 60A subgroup. *Development* **126**, 1753–68.

Sommer, L. (2005). Checkpoints of melanocyte stem cell development. *Sci STKE* **2005**, p. pe42.

Sommer, L., Shah, N., Rao, M., and Anderson, D. J. (1995). The cellular function of MASH1 in autonomic neurogenesis. *Neuron* **15**, pp. 1245–58.

Song, N., Schwab, K. R., Patterson, L. T., Yamaguchi, T., Lin, X., Potter, S. S., and Lang, R. A. (2007). Pygopus 2 has a crucial, Wnt pathway-independent function in lens induction. *Development* **134**, pp. 1873–85.

Sonnenberg-Riethmacher, E., Miehe, M., Stolt, C. C., Goerich, D. E., Wegner, M., and Riethmacher, D. (2001). Development and degeneration of dorsal root ganglia in the absence of the HMG-domain transcription factor Sox10. *Mech Dev* **109**, pp. 253–65.

Soo, K., O'Rourke, M. P., Khoo, P. L., Steiner, K. A., Wong, N., Behringer, R. R., and Tam, P. P. (2002). Twist function is required for the morphogenesis of the cephalic neural tube and the differentiation of the cranial neural crest cells in the mouse embryo. *Dev Biol* **247**, pp. 251–70.

Southard-Smith, E. M., Angrist, M., Ellison, J. S., Agarwala, R., Baxevanis, A. D., Chakravarti, A., and Pavan, W. J. (1999). The Sox10(Dom) mouse: modeling the genetic variation of Waardenburg-Shah (WS4) syndrome. *Genome Res* **9**, pp. 215–25.

Southard-Smith, E. M., Kos, L., and Pavan, W. J. (1998). Sox10 mutation disrupts neural crest development in Dom Hirschsprung mouse model. *Nat Genet* **18**, pp. 60–4.

Sowden, J. C. (2007). Molecular and developmental mechanisms of anterior segment dysgenesis. *Eye (Lond)* **21**, pp. 1310–8.

Sparrow, D. B., Kotecha, S., Towers, N., and Mohun, T. J. (1998). *Xenopus* eHAND: a marker for the developing cardiovascular system of the embryo that is regulated by bone morphogenetic proteins. *Mech Dev* **71**, pp. 151–63.

Spengler, B. A., Lazarova, D. L., Ross, R. A., and Biedler, J. L. (1997). Cell lineage and differentiation state are primary determinants of MYCN gene expression and malignant potential in human neuroblastoma cells. *Oncol Res* **9**, pp. 467–76.

Sperber, S. M., and Dawid, I. B. (2008). barx1 is necessary for ectomesenchyme proliferation and osteochondroprogenitor condensation in the zebrafish pharyngeal arches. *Dev Biol* **321**, pp. 101–10.

Sperber, S. M., Saxena, V., Hatch, G., and Ekker, M. (2008). Zebrafish dlx2a contributes to hindbrain neural crest survival, is necessary for differentiation of sensory ganglia and functions with dlx1a in maturation of the arch cartilage elements. *Dev Biol* **314**, pp. 59–70.

Spinsanti, P., De Vita, T., Caruso, A., Melchiorri, D., Misasi, R., Caricasole, A., and Nicoletti, F. (2008). Differential activation of the calcium/protein kinase C and the canonical beta-catenin pathway by Wnt1 and Wnt7a produces opposite effects on cell proliferation in PC12 cells. *J Neurochem* **104**, pp. 1588–98.

Spokony, R. F., Aoki, Y., Saint-Germain, N., Magner-Fink, E., and Saint-Jeannet, J. P. (2002). The transcription factor Sox9 is required for cranial neural crest development in *Xenopus*. *Development* **129**, pp. 421–32.

Srinivasan, S., Anitha, M., Mwangi, S., and Heuckeroth, R. O. (2005). Enteric neuroblasts require the phosphatidylinositol 3-kinase/Akt/Forkhead pathway for GDNF-stimulated survival. *Mol Cell Neurosci* **29**, pp. 107–19.

Srivastava, D. (1999a). Developmental and genetic aspects of congenital heart disease. *Curr Opin Cardiol* **14**, pp. 263–8.

Srivastava, D. (1999b). HAND proteins: molecular mediators of cardiac development and congenital heart disease. *Trends Cardiovasc Med* **9**, pp. 11–8.

Srivastava, D., Thomas, T., Lin, Q., Kirby, M. L., Brown, D., and Olson, E. N. (1997). Regulation of cardiac mesodermal and neural crest development by the bHLH transcription factor, dHAND. *Nat Genet* **16**, pp. 154–60.

St Amand, T. R., Lu, J. T., and Chien, K. R. (2003). Defects in cardiac conduction system lineages and malignant arrhythmias: developmental pathways and disease. *Novartis Found Symp* **250**, pp. 260–70; discussions 271–5, 276–9.

St Amand, T. R., Lu, J. T., Zamora, M., Gu, Y., Stricker, J., Hoshijima, M., Epstein, J. A., Ross, J. J., Jr., Ruiz-Lozano, P., and Chien, K. R. (2006). Distinct roles of HF-1b/Sp4 in ventricular and neural crest cells lineages affect cardiac conduction system development. *Dev Biol* **291**, pp. 208–17.

Staege, M. S., Hutter, C., Neumann, I., Foja, S., Hattenhorst, U. E., Hansen, G., Afar, D., and Burdach, S. E. (2004). DNA microarrays reveal relationship of Ewing family tumors to both endothelial and fetal neural crest-derived cells and define novel targets. *Cancer Res* **64**, pp. 8213–21.

Stanchina, L., Baral, V., Robert, F., Pingault, V., Lemort, N., Pachnis, V., Goossens, M., and Bondurand, N. (2006). Interactions between Sox10, Edn3 and Ednrb during enteric nervous system and melanocyte development. *Dev Biol* **295**, pp. 232–49.

Steiner, A. B., Engleka, M. J., Lu, Q., Piwarzyk, E. C., Yaklichkin, S., Lefebvre, J. L., Walters, J. W., Pineda-Salgado, L., Labosky, P. A., and Kessler, D. S. (2006). FoxD3 regulation of Nodal in the Spemann organizer is essential for *Xenopus* dorsal mesoderm development. *Development* **133**, pp. 4827–38.

Steingrimsson, E., Moore, K. J., Lamoreux, M. L., Ferre-D'Amare, A. R., Burley, S. K., Zimring, D. C., Skow, L. C., Hodgkinson, C. A., Arnheiter, H., Copeland, N. G., and et al. (1994). Molecular basis of mouse microphthalmia (mi) mutations helps explain their developmental and phenotypic consequences. *Nat Genet* **8**, pp. 256–63.

Steventon, B., Carmona-Fontaine, C., and Mayor, R. (2005). Genetic network during neural crest induction: from cell specification to cell survival. *Semin Cell Dev Biol* **16**, pp. 647–54.

Stewart, R. A., Arduini, B. L., Berghmans, S., George, R. E., Kanki, J. P., Henion, P. D., and Look, A. T. (2006). Zebrafish foxd3 is selectively required for neural crest specification, migration and survival. *Dev Biol* **292**, pp. 174–88.

Stoetzel, C., Weber, B., Bourgeois, P., Bolcato-Bellemin, A. L., and Perrin-Schmitt, F. (1995). Dorso-ventral and rostro-caudal sequential expression of M-twist in the postimplantation murine embryo. *Mech Dev* **51**, pp. 251–63.

Stolt, C. C., Lommes, P., Hillgartner, S., and Wegner, M. (2008). The transcription factor Sox5 modulates Sox10 function during melanocyte development. *Nucleic Acids Res* **36**, pp. 5427–40.

Stolt, C. C., Rehberg, S., Ader, M., Lommes, P., Riethmacher, D., Schachner, M., Bartsch, U., and

Wegner, M. (2002). Terminal differentiation of myelin-forming oligodendrocytes depends on the transcription factor Sox10. *Genes Dev* **16**, pp. 165–70.

Studer, M., Popperl, H., Marshall, H., Kuroiwa, A., and Krumlauf, R. (1994). Role of a conserved retinoic acid response element in rhombomere restriction of Hoxb-1. *Science* **265**, pp. 1728–32.

Sucov, H. M., Dyson, E., Gumeringer, C. L., Price, J., Chien, K. R., and Evans, R. M. (1994). RXR alpha mutant mice establish a genetic basis for vitamin A signaling in heart morphogenesis. *Genes Dev* **8**, pp. 1007–18.

Sun, C. S., Wu, K. T., Lee, H. H., Uen, Y. H., Tian, Y. F., Tzeng, C. C., Wang, A. H., Cheng, C. J., and Tsai, S. L. (2008). Anti-sense morpholino oligonucleotide assay shows critical involvement for NF-kappaB activation in the production of Wnt-1 protein by HepG2 cells: oncology implications. *J Biomed Sci* **15**, pp. 633–43.

Sun Rhodes, L. S., and Merzdorf, C. S. (2006). The zic1 gene is expressed in chick somites but not in migratory neural crest. *Gene Expr Patterns* **6**, pp. 539–45.

Sun, S. N., Gui, Y. H., Wang, Y. X., Qian, L. X., Jiang, Q., Liu, D., and Song, H. Y. (2007). Effect of dihydrofolate reductase gene knock-down on the expression of heart and neural crest derivatives expressed transcript 2 in zebrafish cardiac development. *Chin Med J (Engl)* **120**, pp. 1166–71.

Suzuki, T., Sakai, D., Osumi, N., Wada, H., and Wakamatsu, Y. (2006). Sox genes regulate type 2 collagen expression in avian neural crest cells. *Dev Growth Differ* **48**, pp. 477–86.

Svaren, J., and Meijer, D. (2008). The molecular machinery of myelin gene transcription in Schwann cells. *Glia* **56**, pp. 1541–51.

Sznajer, Y., Coldea, C., Meire, F., Delpierre, I., Sekhara, T., and Touraine, R. L. (2008). A de novo SOX10 mutation causing severe type 4 Waardenburg syndrome without Hirschsprung disease. *Am J Med Genet A* **146A**, pp. 1038–41.

Sztriha, L., Espinosa-Parrilla, Y., Gururaj, A., Amiel, J., Lyonnet, S., Gerami, S., and Johansen, J. G. (2003). Frameshift mutation of the zinc finger homeo box 1 B gene in syndromic corpus callosum agenesis (Mowat–Wilson syndrome). *Neuropediatrics* **34**, pp. 322–5.

Tachibana, M. (2000). MITF: a stream flowing for pigment cells. *Pigment Cell Res* **13**, pp. 230–40.

Tachibana, M., Kobayashi, Y., and Matsushima, Y. (2003). Mouse models for four types of Waardenburg syndrome. *Pigment Cell Res* **16**, pp. 448–54.

Tahtakran, S. A., and Selleck, M. A. (2003). Ets-1 expression is associated with cranial neural crest migration and vasculogenesis in the chick embryo. *Gene Expr Patterns* **3**, pp. 455–8.

Takagi, T., Moribe, H., Kondoh, H., and Higashi, Y. (1998). DeltaEF1, a zinc finger and homeodomain transcription factor, is required for skeleton patterning in multiple lineages. *Development* **125**, pp. 21–31.

Takahashi, E., Funato, N., Higashihori, N., Hata, Y., Gridley, T., and Nakamura, M. (2004). Snail regulates p21(WAF/CIP1) expression in cooperation with E2A and Twist. *Biochem Biophys Res Commun* **325**, pp. 1136–44.

Takahashi, K., Nuckolls, G. H., Takahashi, I., Nonaka, K., Nagata, M., Ikura, T., Slavkin, H. C., and Shum, L. (2001). Msx2 is a repressor of chondrogenic differentiation in migratory cranial neural crest cells. *Dev Dyn* **222**, pp. 252–62.

Takahashi, K., Tanabe, K., Ohnuki, M., Narita, M., Ichisaka, T., Tomoda, K., and Yamanaka, S. (2007). Induction of pluripotent stem cells from adult human fibroblasts by defined factors. *Cell* **131**, pp. 861–72.

Takahashi, K., and Yamanaka, S. (2006). Induction of pluripotent stem cells from mouse embryonic and adult fibroblast cultures by defined factors. *Cell* **126**, pp. 663–76.

Takeda, K., Yasumoto, K., Takada, R., Takada, S., Watanabe, K., Udono, T., Saito, H., Takahashi, K., and Shibahara, S. (2000). Induction of melanocyte-specific microphthalmia-associated transcription factor by Wnt-3a. *J Biol Chem* **275**, pp. 14013–6.

Takeuchi, J. K., and Bruneau, B. G. (2009). Directed transdifferentiation of mouse mesoderm to heart tissue by defined factors. *Nature* **459**, pp. 708–11.

Talbot, D., Loring, J., and Schorle, H. (1999). Spatiotemporal expression pattern of keratins in skin of AP-2alpha-deficient mice. *J Invest Dermatol* **113**, pp. 816–20.

Taneyhill, L. A., and Bronner-Fraser, M. (2005). Dynamic alterations in gene expression after Wnt-mediated induction of avian neural crest. *Mol Biol Cell* **16**, pp. 5283–93.

Taneyhill, L. A., Coles, E. G., and Bronner-Fraser, M. (2007). Snail2 directly represses cadherin6B during epithelial-to-mesenchymal transitions of the neural crest. *Development* **134**, pp. 1481–90.

Tang, L. S., Wlodarczyk, B. J., Santillano, D. R., Miranda, R. C., and Finnell, R. H. (2004). Developmental consequences of abnormal folate transport during murine heart morphogenesis. *Birth Defects Res A Clin Mol Teratol* **70**, pp. 449–58.

Tanno, B., Negroni, A., Vitali, R., Pirozzoli, M. C., Cesi, V., Mancini, C., Calabretta, B., and Raschella, G. (2002). Expression of insulin-like growth factor-binding protein 5 in neuroblastoma cells is regulated at the transcriptional level by c-Myb and B-Myb via direct and indirect mechanisms. *J Biol Chem* **277**, pp. 23172–80.

Tassabehji, M., Newton, V. E., Leverton, K., Turnbull, K., Seemanova, E., Kunze, J., Sperling, K., Strachan, T., and Read, A. P. (1994). PAX3 gene structure and mutations: close analogies between Waardenburg syndrome and the Splotch mouse. *Hum Mol Genet* **3**, pp. 1069–74.

Tassabehji, M., Newton, V. E., Liu, X. Z., Brady, A., Donnai, D., Krajewska-Walasek, M., Murday, V., Norman, A., Obersztyn, E., Reardon, W., and et al. (1995). The mutational spectrum in Waardenburg syndrome. *Hum Mol Genet* **4**, pp. 2131–7.

Tassabehji, M., Read, A. P., Newton, V. E., Patton, M., Gruss, P., Harris, R., and Strachan, T. (1993). Mutations in the PAX3 gene causing Waardenburg syndrome type 1 and type 2. *Nat Genet* **3**, pp. 26–30.

Taylor, K. M., and Labonne, C. (2005). SoxE factors function equivalently during neural crest and inner ear development and their activity is regulated by SUMOylation. *Dev Cell* **9**, pp. 593–603.

Tazi, A., Czernichow, P., and Scharfmann, R. (1995). Similarities and discrepancies in the signaling pathway for nerve growth factor in an insulin producing cell line and a neural crest-derived cell line. *J Neuroendocrinol* **7**, pp. 29–36.

Techawattanawisal, W., Nakahama, K., Komaki, M., Abe, M., Takagi, Y., and Morita, I. (2007). Isolation of multipotent stem cells from adult rat periodontal ligament by neurosphere-forming culture system. *Biochem Biophys Res Commun* **357**, pp. 917–23.

Teng, L., and Labosky, P. A. (2006). Neural crest stem cells. *Adv Exp Med Biol* **589**, pp. 206–12.

Teng, L., Mundell, N. A., Frist, A. Y., Wang, Q., and Labosky, P. A. (2008). Requirement for Foxd3 in the maintenance of neural crest progenitors. *Development* **135**, pp. 1615–24.

Terui, E., Matsunaga, T., Yoshida, H., Kouchi, K., Kuroda, H., Hishiki, T., Saito, T., Yamada, S., Shirasawa, H., and Ohnuma, N. (2005). Shc family expression in neuroblastoma: high expression of shcC is associated with a poor prognosis in advanced neuroblastoma. *Clin Cancer Res* **11**, pp. 3280–7.

Theveneau, E., Duband, J. L., and Altabef, M. (2007). Ets-1 confers cranial features on neural crest delamination. *PLoS One* **2**, p. e1142.

Thiele, C. J., Deutsch, L. A., and Israel, M. A. (1988). The expression of multiple proto-oncogenes is differentially regulated during retinoic acid induced maturation of human neuroblastoma cell lines. *Oncogene* **3**, pp. 281–8.

Thisse, C., Thisse, B., and Postlethwait, J. H. (1995). Expression of snail2, a second member of the zebrafish snail family, in cephalic mesendoderm and presumptive neural crest of wild-type and spadetail mutant embryos. *Dev Biol* **172**, pp. 86–99.

Thomas, A. J., and Erickson, C. A. (2009). FOXD3 regulates the lineage switch between neural crest-derived glial cells and pigment cells by repressing MITF through a non-canonical mechanism. *Development* **136**, pp. 1849–58.

Thomas, B. L., Tucker, A. S., Qui, M., Ferguson, C. A., Hardcastle, Z., Rubenstein, J. L., and Sharpe, P. T. (1997). Role of Dlx-1 and Dlx-2 genes in patterning of the murine dentition. *Development* **124**, pp. 4811–8.

Thomas, S., Thomas, M., Wincker, P., Babarit, C., Xu, P., Speer, M. C., Munnich, A., Lyonnet, S., Vekemans, M., and Etchevers, H. C. (2008). Human neural crest cells display molecular and phenotypic hallmarks of stem cells. *Hum Mol Genet* **17**, pp. 3411–25.

Thomas, T., Kurihara, H., Yamagishi, H., Kurihara, Y., Yazaki, Y., Olson, E. N., and Srivastava, D. (1998). A signaling cascade involving endothelin-1, dHAND and msx1 regulates development of neural-crest-derived branchial arch mesenchyme. *Development* **125**, pp. 3005–14.

Thomas, W. D., Raif, A., Hansford, L., and Marshall, G. (2004). N-myc transcription molecule and oncoprotein. *Int J Biochem Cell Biol* **36**, pp. 771–5.

Thomson, J. A., Murphy, K., Baker, E., Sutherland, G. R., Parsons, P. G., Sturm, R. A., and Thomson, F. (1995). The brn-2 gene regulates the melanocytic phenotype and tumorigenic potential of human melanoma cells. *Oncogene* **11**, pp. 691–700.

Ting, M. C., Wu, N. L., Roybal, P. G., Sun, J., Liu, L., Yen, Y., and Maxson, R. E., Jr. (2009). EphA4 as an effector of Twist1 in the guidance of osteogenic precursor cells during calvarial bone growth and in craniosynostosis. *Development* **136**, pp. 855–64.

Tissier-Seta, J. P., Mucchielli, M. L., Mark, M., Mattei, M. G., Goridis, C., and Brunet, J. F. (1995). Barx1, a new mouse homeodomain transcription factor expressed in cranio-facial ectomesenchyme and the stomach. *Mech Dev* **51**, pp. 3–15.

Toma, J. G., McKenzie, I. A., Bagli, D., and Miller, F. D. (2005). Isolation and characterization of multipotent skin-derived precursors from human skin. *Stem Cells* **23**, pp. 727–37.

Torlakovic, E. E., Bilalovic, N., Nesland, J. M., Torlakovic, G., and Florenes, V. A. (2004). Ets-1 transcription factor is widely expressed in benign and malignant melanocytes and its expression has no significant association with prognosis. *Mod Pathol* **17**, pp. 1400–6.

Touraine, R. L., Attie-Bitach, T., Manceau, E., Korsch, E., Sarda, P., Pingault, V., Encha-Razavi, F., Pelet, A., Auge, J., Nivelon-Chevallier, A., Holschneider, A. M., Munnes, M., Doerfler, W., Goossens, M., Munnich, A., Vekemans, M., and Lyonnet, S. (2000). Neurological phenotype in Waardenburg syndrome type 4 correlates with novel SOX10 truncating mutations and expression in developing brain. *Am J Hum Genet* **66**, pp. 1496–503.

Tremblay, P., Kessel, M., and Gruss, P. (1995). A transgenic neuroanatomical marker identifies cranial neural crest deficiencies associated with the Pax3 mutant Splotch. *Dev Biol* **171**, pp. 317–29.

Tribulo, C., Aybar, M. J., Nguyen, V. H., Mullins, M. C., and Mayor, R. (2003). Regulation of Msx genes by a Bmp gradient is essential for neural crest specification. *Development* **130**, pp. 6441–52.

Tribulo, C., Aybar, M. J., Sanchez, S. S., and Mayor, R. (2004). A balance between the anti-apoptotic activity of Slug and the apoptotic activity of msx1 is required for the proper development of the neural crest. *Dev Biol* **275**, pp. 325–42.

Tropepe, V., Li, S., Dickinson, A., Gamse, J. T., and Sive, H. L. (2006). Identification of a BMP inhibitor-responsive promoter module required for expression of the early neural gene zic1. *Dev Biol* **289**, pp. 517–29.

Tsarovina, K., Pattyn, A., Stubbusch, J., Muller, F., van der Wees, J., Schneider, C., Brunet, J. F., and Rohrer, H. (2004). Essential role of Gata transcription factors in sympathetic neuron development. *Development* **131**, pp. 4775–86.

Tucker, A. S., Yamada, G., Grigoriou, M., Pachnis, V., and Sharpe, P. T. (1999). Fgf-8 determines rostral-caudal polarity in the first branchial arch. *Development* **126**, pp. 51–61.

Tucker, R. P. (2004). Neural crest cells: a model for invasive behavior. *Int J Biochem Cell Biol* **36**, pp. 173–7.

Twigg, S. R., and Wilkie, A. O. (1999). Characterisation of the human snail (SNAI1) gene and exclusion as a major disease gene in craniosynostosis. *Hum Genet* **105**, pp. 320–6.

Uchikawa, M., Ishida, Y., Takemoto, T., Kamachi, Y., and Kondoh, H. (2003). Functional analysis of chicken Sox2 enhancers highlights an array of diverse regulatory elements that are conserved in mammals. *Dev Cell* **4**, pp. 509–19.

Unsicker, K., Huber, K., Schutz, G., and Kalcheim, C. (2005). The chromaffin cell and its development. *Neurochem Res* **30**, pp. 921–5.

Vallin, J., Thuret, R., Giacomello, E., Faraldo, M. M., Thiery, J. P., and Broders, F. (2001). Cloning and characterization of three *Xenopus* slug promoters reveal direct regulation by Lef/beta-catenin signaling. *J Biol Chem* **276**, pp. 30350–8.

Van Camp, G., Van Thienen, M. N., Handig, I., Van Roy, B., Rao, V. S., Milunsky, A., Read, A. P., Baldwin, C. T., Farrer, L. A., Bonduelle, M., and et al. (1995). Chromosome 13q deletion with Waardenburg syndrome: further evidence for a gene involved in neural crest function on 13q. *J Med Genet* **32**, pp. 531–6.

van Grunsven, L. A., Papin, C., Avalosse, B., Opdecamp, K., Huylebroeck, D., Smith, J. C., and Bellefroid, E. J. (2000). XSIP1, a *Xenopus* zinc finger/homeodomain encoding gene highly expressed during early neural development. *Mech Dev* **94**, pp. 189–93.

van Grunsven, L. A., Taelman, V., Michiels, C., Opdecamp, K., Huylebroeck, D., and Bellefroid, E. J. (2006). deltaEF1 and SIP1 are differentially expressed and have overlapping activities during *Xenopus* embryogenesis. *Dev Dyn* **235**, pp. 1491–500.

Veenstra, G. J., Beumer, T. L., Peterson-Maduro, J., Stegeman, B. I., Karg, H. A., van der Vliet, P. C., and Destree, O. H. (1995). Dynamic and differential Oct-1 expression during early *Xenopus* embryogenesis: persistence of Oct-1 protein following down-regulation of the RNA. *Mech Dev* **50**, pp. 103–17.

Verheij, J. B., Sival, D. A., van der Hoeven, J. H., Vos, Y. J., Meiners, L. C., Brouwer, O. F., and van Essen, A. J. (2006). Shah-Waardenburg syndrome and PCWH associated with SOX10 mutations: a case report and review of the literature. *Eur J Paediatr Neurol* **10**, pp. 11–7.

Verzi, M. P., Agarwal, P., Brown, C., McCulley, D. J., Schwarz, J. J., and Black, B. L. (2007).

The transcription factor MEF2C is required for craniofacial development. *Dev Cell* **12**, pp. 645–52.

Villanueva, M. P., Aiyer, A. R., Muller, S., Pletcher, M. T., Liu, X., Emanuel, B., Srivastava, D., and Reeves, R. H. (2002). Genetic and comparative mapping of genes dysregulated in mouse hearts lacking the Hand2 transcription factor gene. *Genomics* **80**, pp. 593–600.

Vincent, S. D., Dunn, N. R., Sciammas, R., Shapiro-Shalef, M., Davis, M. M., Calame, K., Bikoff, E. K., and Robertson, E. J. (2005). The zinc finger transcriptional repressor Blimp1/Prdm1 is dispensable for early axis formation but is required for specification of primordial germ cells in the mouse. *Development* **132**, pp. 1315–25.

Vincentz, J. W., Barnes, R. M., Rodgers, R., Firulli, B. A., Conway, S. J., and Firulli, A. B. (2008). An absence of Twist1 results in aberrant cardiac neural crest morphogenesis. *Dev Biol* **320**, pp. 131–9.

Vlaeminck-Guillem, V., Carrere, S., Dewitte, F., Stehelin, D., Desbiens, X., and Duterque-Coquillaud, M. (2000). The Ets family member Erg gene is expressed in mesodermal tissues and neural crests at fundamental steps during mouse embryogenesis. *Mech Dev* **91**, pp. 331–5.

Voiculescu, O., Taillebourg, E., Pujades, C., Kress, C., Buart, S., Charnay, P., and Schneider-Maunoury, S. (2001). Hindbrain patterning: Krox20 couples segmentation and specification of regional identity. *Development* **128**, pp. 4967–78.

Wada, N., Javidan, Y., Nelson, S., Carney, T. J., Kelsh, R. N., and Schilling, T. F. (2005). Hedgehog signaling is required for cranial neural crest morphogenesis and chondrogenesis at the midline in the zebrafish skull. *Development* **132**, pp. 3977–88.

Wada, R. K., Pai, D. S., Huang, J., Yamashiro, J. M., and Sidell, N. (1997). Interferon-gamma and retinoic acid down-regulate N-myc in neuroblastoma through complementary mechanisms of action. *Cancer Lett* **121**, pp. 181–8.

Wakamatsu, Y., Endo, Y., Osumi, N., and Weston, J. A. (2004). Multiple roles of Sox2, an HMG-box transcription factor in avian neural crest development. *Dev Dyn* **229**, pp. 74–86.

Wakamatsu, Y., Watanabe, Y., Nakamura, H., and Kondoh, H. (1997). Regulation of the neural crest cell fate by N-myc: promotion of ventral migration and neuronal differentiation. *Development* **124**, pp. 1953–62.

Wang, Q., Kumar, S., Mitsios, N., Slevin, M., and Kumar, P. (2007). Investigation of downstream target genes of PAX3c, PAX3e and PAX3g isoforms in melanocytes by microarray analysis. *Int J Cancer* **120**, pp. 1223–31.

Wang, Q., Kumar, S., Slevin, M., and Kumar, P. (2006). Functional analysis of alternative isoforms of the transcription factor PAX3 in melanocytes in vitro. *Cancer Res* **66**, pp. 8574–80.

Wang, W., Lo, P., Frasch, M., and Lufkin, T. (2000). Hmx: an evolutionary conserved homeobox gene family expressed in the developing nervous system in mice and *Drosophila*. *Mech Dev* **99**, pp. 123–37.

Warner, S. J., Hutson, M. R., Oh, S. H., Gerlach-Bank, L. M., Lomax, M. I., and Barald, K. F. (2003). Expression of ZIC genes in the development of the chick inner ear and nervous system. *Dev Dyn* **226**, pp. 702–12.

Watanabe, K., Takeda, K., Yasumoto, K., Udono, T., Saito, H., Ikeda, K., Takasaka, T., Takahashi, K., Kobayashi, T., Tachibana, M., and Shibahara, S. (2002a). Identification of a distal enhancer for the melanocyte-specific promoter of the MITF gene. *Pigment Cell Res* **15**, pp. 201–11.

Watanabe, M., Layne, M. D., Hsieh, C. M., Maemura, K., Gray, S., Lee, M. E., and Jain, M. K. (2002b). Regulation of smooth muscle cell differentiation by AT-rich interaction domain transcription factors Mrf2alpha and Mrf2beta. *Circ Res* **91**, pp. 382–9.

Wei, K., Chen, J., Akrami, K., Sekhon, R., and Chen, F. (2007). Generation of mice deficient for Lbx2, a gene expressed in the urogenital system, nervous system, and Pax3 dependent tissues. *Genesis* **45**, pp. 361–8.

Weiss, K., Stock, D., Zhao, Z., Buchanan, A., Ruddle, F., and Shashikant, C. (1998). Perspectives on genetic aspects of dental patterning. *Eur J Oral Sci* **106 Suppl 1**, pp. 55–63.

Weninger, W. J., Lopes Floro, K., Bennett, M. B., Withington, S. L., Preis, J. I., Barbera, J. P., Mohun, T. J., and Dunwoodie, S. L. (2005). Cited2 is required both for heart morphogenesis and establishment of the left–right axis in mouse development. *Development* **132**, pp. 1337–48.

Werling, U., and Schorle, H. (2002). Transcription factor gene AP-2 gamma essential for early murine development. *Mol Cell Biol* **22**, pp. 3149–56.

Werner, T., Hammer, A., Wahlbuhl, M., Bosl, M. R., and Wegner, M. (2007). Multiple conserved regulatory elements with overlapping functions determine Sox10 expression in mouse embryogenesis. *Nucleic Acids Res* **35**, pp. 6526–38.

Whitlock, K. E., Smith, K. M., Kim, H., and Harden, M. V. (2005). A role for foxd3 and sox10 in the differentiation of gonadotropin-releasing hormone (GnRH) cells in the zebrafish Danio rerio. *Development* **132**, pp. 5491–502.

Widera, D., Zander, C., Heidbreder, M., Kasperek, Y., Noll, T., Seitz, O., Saldamli, B., Sudhoff, H., Sader, R., Kaltschmidt, C., and Kaltschmidt, B. (2009). Adult palatum as a novel source of neural crest-related stem cells. *Stem Cells* **27**, pp. 1899–910.

Widlund, H. R., and Fisher, D. E. (2003). Microphthalamia-associated transcription factor: a critical regulator of pigment cell development and survival. *Oncogene* **22**, pp. 3035–41.

Widlund, H. R., Horstmann, M. A., Price, E. R., Cui, J., Lessnick, S. L., Wu, M., He, X., and Fisher, D. E. (2002). Beta-catenin-induced melanoma growth requires the downstream target Microphthalmia-associated transcription factor. *J Cell Biol* **158**, pp. 1079–87.

Wilkinson, D. G. (1993). Molecular mechanisms of segmental patterning in the vertebrate hindbrain. *Perspect Dev Neurobiol* **1**, pp. 117–25.

Wilkinson, D. G., Bhatt, S., Chavrier, P., Bravo, R., and Charnay, P. (1989). Segment-specific expression of a zinc-finger gene in the developing nervous system of the mouse. *Nature* **337**, pp. 461–4.

Williams, J. A., Mann, F. M., and Brown, N. A. (1997). Gene expression domains as markers in developmental toxicity studies using mammalian embryo culture. *Int J Dev Biol* **41**, pp. 359–64.

Williams, T., and Tjian, R. (1991). Analysis of the DNA-binding and activation properties of the human transcription factor AP-2. *Genes Dev* **5**, pp. 670–82.

Wlodraczyk, B., Bennett, G. D., Calvin, J. A., Craig, J. C., and Finnell, R. H. (1996). Arsenic-induced alterations in embryonic transcription factor gene expression: implications for abnormal neural development. *Dev Genet* **18**, pp. 306–15.

Wong, C. E., Paratore, C., Dours-Zimmermann, M. T., Rochat, A., Pietri, T., Suter, U., Zimmermann, D. R., Dufour, S., Thiery, J. P., Meijer, D., Beermann, F., Barrandon, Y., and Sommer, L. (2006). Neural crest-derived cells with stem cell features can be traced back to multiple lineages in the adult skin. *J Cell Biol* **175**, pp. 1005–15.

Wood, J. N. (1995). Regulation of NF-kappa B activity in rat dorsal root ganglia and PC12 cells by tumour necrosis factor and nerve growth factor. *Neurosci Lett* **192**, pp. 41–4.

Woodside, K. J., Shen, H., Muntzel, C., Daller, J. A., Sommers, C. L., and Love, P. E. (2004). Expression of Dlx and Lhx family homeobox genes in fetal thymus and thymocytes. *Gene Expr Patterns* **4**, pp. 315–20.

Wu, M., Li, J., Engleka, K. A., Zhou, B., Lu, M. M., Plotkin, J. B., and Epstein, J. A. (2008). Persistent expression of Pax3 in the neural crest causes cleft palate and defective osteogenesis in mice. *J Clin Invest* **118**, pp. 2076–87.

Wu, X., and Howard, M. J. (2001). Two signal transduction pathways involved in the catecholaminergic differentiation of avian neural crest-derived cells in vitro. *Mol Cell Neurosci* **18**, pp. 394–406.

Wu, X., and Howard, M. J. (2002). Transcripts encoding HAND genes are differentially expressed and regulated by BMP4 and GDNF in developing avian gut. *Gene Expr* **10**, pp. 279–93.

Xu, J., Watts, J. A., Pope, S. D., Gadue, P., Kamps, M., Plath, K., Zaret, K. S., and Smale, S. T. (2009). Transcriptional competence and the active marking of tissue-specific enhancers by defined transcription factors in embryonic and induced pluripotent stem cells. *Genes Dev* **23**, pp. 2824–38.

Xu, X., Jeong, L., Han, J., Ito, Y., Bringas, P., Jr., and Chai, Y. (2003). Developmental expression of Smad1–7 suggests critical function of TGF-beta/BMP signaling in regulating epithelial–mesenchymal interaction during tooth morphogenesis. *Int J Dev Biol* **47**, pp. 31–9.

Ya, J., Schilham, M. W., de Boer, P. A., Moorman, A. F., Clevers, H., and Lamers, W. H. (1998). Sox4-deficiency syndrome in mice is an animal model for common trunk. *Circ Res* **83**, pp. 986–94.

Yajima, I., Sato, S., Kimura, T., Yasumoto, K., Shibahara, S., Goding, C. R., and Yamamoto, H. (1999). An L1 element intronic insertion in the black-eyed white (Mitf[mi-bw]) gene: the loss of a single Mitf isoform responsible for the pigmentary defect and inner ear deafness. *Hum Mol Genet* **8**, pp. 1431–41.

Yaklichkin, S., Steiner, A. B., Lu, Q., and Kessler, D. S. (2007). FoxD3 and Grg4 physically interact to repress transcription and induce mesoderm in *Xenopus*. *J Biol Chem* **282**, pp. 2548–57.

Yamagata, M., and Noda, M. (1998). The winged-helix transcription factor CWH-3 is expressed in developing neural crest cells. *Neurosci Lett* **249**, pp. 33–6.

Yamagishi, H., Garg, V., Matsuoka, R., Thomas, T., and Srivastava, D. (1999). A molecular pathway revealing a genetic basis for human cardiac and craniofacial defects. *Science* **283**, pp. 1158–61.

Yamamoto, M., Watt, C. D., Schmidt, R. J., Kuscuoglu, U., Miesfeld, R. L., and Goldhamer, D. J. (2007). Cloning and characterization of a novel MyoD enhancer-binding factor. *Mech Dev* **124**, pp. 715–28.

Yan, Y. L., Jowett, T., and Postlethwait, J. H. (1998). Ectopic expression of hoxb2 after retinoic acid treatment or mRNA injection: disruption of hindbrain and craniofacial morphogenesis in zebrafish embryos. *Dev Dyn* **213**, pp. 370–85.

Yan, Y. L., Miller, C. T., Nissen, R. M., Singer, A., Liu, D., Kirn, A., Draper, B., Willoughby, J., Morcos, P. A., Amsterdam, A., Chung, B. C., Westerfield, M., Haffter, P., Hopkins, N., Kimmel, C., and Postlethwait, J. H. (2002). A zebrafish sox9 gene required for cartilage morphogenesis. *Development* **129**, pp. 5065–79.

Yan, Y. L., Willoughby, J., Liu, D., Crump, J. G., Wilson, C., Miller, C. T., Singer, A., Kimmel, C., Westerfield, M., and Postlethwait, J. H. (2005). A pair of Sox: distinct and overlapping functions of zebrafish sox9 co-orthologs in craniofacial and pectoral fin development. *Development* **132**, pp. 1069–83.

Yang, L., Zhang, H., Hu, G., Wang, H., Abate-Shen, C., and Shen, M. M. (1998). An early phase of embryonic Dlx5 expression defines the rostral boundary of the neural plate. *J Neurosci* **18**, pp. 8322–30.

Yasumoto, K., Takeda, K., Saito, H., Watanabe, K., Takahashi, K., and Shibahara, S. (2002). Microphthalmia-associated transcription factor interacts with LEF-1, a mediator of Wnt signaling. *Embo J* **21**, pp. 2703–14.

Ye, M., Coldren, C., Liang, X., Mattina, T., Goldmuntz, E., Benson, D. W., Ivy, D., Perryman, M. B., Garrett-Sinha, L. A., and Grossfeld, P. Deletion of ETS-1, a gene in the Jacobsen syndrome critical region, causes ventricular septal defects and abnormal ventricular morphology in mice. *Hum Mol Genet* **19**, 648–56.

Yeo, G. H., Cheah, F. S., Jabs, E. W., and Chong, S. S. (2007). Zebrafish twist1 is expressed in craniofacial, vertebral, and renal precursors. *Dev Genes Evol* **217**, pp. 783–9.

Young, H. M., Bergner, A. J., Anderson, R. B., Enomoto, H., Milbrandt, J., Newgreen, D. F., and Whitington, P. M. (2004). Dynamics of neural crest-derived cell migration in the embryonic mouse gut. *Dev Biol* **270**, pp. 455–73.

Young, H. M., Bergner, A. J., and Muller, T. (2003). Acquisition of neuronal and glial markers by neural crest-derived cells in the mouse intestine. *J Comp Neurol* **456**, pp. 1–11.

Yu, H., Fang, D., Kumar, S. M., Li, L., Nguyen, T. K., Acs, G., Herlyn, M., and Xu, X. (2006). Isolation of a novel population of multipotent adult stem cells from human hair follicles. *Am J Pathol* **168**, pp. 1879–88.

Yu, J., Hu, K., Smuga-Otto, K., Tian, S., Stewart, R., Slukvin, II, and Thomson, J. A. (2009). Human induced pluripotent stem cells free of vector and transgene sequences. *Science* **324**, pp. 797–801.

Zaret, K. S., Watts, J., Xu, J., Wandzioch, E., Smale, S. T., and Sekiya, T. (2008). Pioneer factors, genetic competence, and inductive signaling: programming liver and pancreas progenitors from the endoderm. *Cold Spring Harb Symp Quant Biol* **73**, pp. 119–26.

Zhang, C., Carl, T. F., Trudeau, E. D., Simmet, T., and Klymkowsky, M. W. (2006a). An NF-kappaB and slug regulatory loop active in early vertebrate mesoderm. *PLoS One* **1**, p. e106.

Zhang, F., Nagy Kovacs, E., and Featherstone, M. S. (2000). Murine hoxd4 expression in the CNS requires multiple elements including a retinoic acid response element. *Mech Dev* **96**, pp. 79–89.

Zhang, J., Hagopian-Donaldson, S., Serbedzija, G., Elsemore, J., Plehn-Dujowich, D., McMahon, A. P., Flavell, R. A., and Williams, T. (1996). Neural tube, skeletal and body wall defects in mice lacking transcription factor AP-2. *Nature* **381**, pp. 238–41.

Zhang, Y., Luo, T., and Sargent, T. D. (2006b). Expression of TFAP2beta and TFAP2gamma genes in *Xenopus laevis*. *Gene Expr Patterns* **6**, pp. 589–95.

Zhao, M., Isom, S. C., Lin, H., Hao, Y., Zhang, Y., Zhao, J., Whyte, J. J., Dobbs, K. B., and Prather, R. S. (2009). Tracing the stemness of porcine skin-derived progenitors (pSKP) back to specific marker gene expression. *Cloning Stem Cells* **11**, pp. 111–22.

Zhou, H. M., Wang, J., Rogers, R., and Conway, S. J. (2008a). Lineage-specific responses to reduced embryonic Pax3 expression levels. *Dev Biol* **315**, pp. 369–82.

Zhou, Q., Brown, J., Kanarek, A., Rajagopal, J., and Melton, D. A. (2008b). In vivo reprogramming of adult pancreatic exocrine cells to beta-cells. *Nature* **455**, pp. 627–32.

Zhu, L., Lee, H. O., Jordan, C. S., Cantrell, V. A., Southard-Smith, E. M., and Shin, M. K. (2004). Spatiotemporal regulation of endothelin receptor-B by SOX10 in neural crest-derived enteric neuron precursors. *Nat Genet* **36**, pp. 732–7.

Zhu, L., Peng, J. L., Harutyunyan, K. G., Garcia, M. D., Justice, M. J., and Belmont, J. W. (2007). Craniofacial, skeletal, and cardiac defects associated with altered embryonic murine Zic3 expression following targeted insertion of a PGK-NEO cassette. *Front Biosci* **12**, pp. 1680–90.

Author Biographies

Patricia A. Labosky was born in Niagara Falls, NY. Trish grew up and was educated in Youngstown, NY. She received her BA in Biology from the University of Pennsylvania in 1985 and her PhD in 1992 in the Department of Biology at Wesleyan University in Middletown, CT, where she worked with Dr. Laura Grabel on the role of homeobox transcription factors in differentiation of embryonal carcinoma (EC) cells *in vitro*. Dr. Labosky joined the laboratory of Dr. Brigid L.M. Hogan at Vanderbilt University as a postdoctoral fellow. Her work focused on the derivation of stem cells from primordial germ cells and the role of winged helix genes in early embryogenesis. After accepting an appointment as an Assistant Professor in the Department of Cell and Developmental Biology at the University of Pennsylvania in 1997, Dr. Labosky continued her work on the role of winged helix genes in patterning the vertebrate embryo. She moved to Vanderbilt University as an Associate Professor in 2006 and has focused on molecular mechanisms required to maintain pluripotent embryonic stem (ES) cells and neural crest cell development in the mouse.

Brian L. Nelms grew up in Lancaster County, Pennsylvania. He received his bachelor's degrees in Biochemistry and Molecular Biology and in Spanish from Pennsylvania State University. He continued his graduate education at Pennsylvania State University, receiving his PhD from the Department of Biochemistry and Molecular Biology in the laboratory of Dr. Wendy Hanna-Rose for his studies on transcriptional regulation of the development of the *Caenorhabditis elegans* egg-laying apparatus. After completing his PhD studies, Dr. Nelms joined the laboratory of Dr. Patricia A. Labosky at Vanderbilt University, where he began studies in transcriptional control of cardiac neural crest development in mouse and pursuing his long-held interest in the fields of developmental biology and transcriptional regulation.

